Obstacles to Mineral Development

(Pergamon Policy Studies-18)

Pergamon Titles of Related Interest

Barney *The Global 2000 Report to the President of the U.S.: Entering the 21st Century*

Diwan/Livingston *Alternative Development Strategies and Appropriate Technology: Science Policy for an Equitable World Order*

Godet *The Crisis in Forecasting and the Emergence of the "Prospective" Approach: With Case Studies in Energy and Air Transport*

Goodman/Love *Management of Development Projects: An International Case Study Approach*

Menon *Bridges Across the South: Technical Cooperation Among Developing Countries*

The Rothko Chapel *Toward a New Strategy for Development*

Thomas/Wionczek *Integration of Science and Technology with Development: Caribbean and Latin American Problems in the Context of the United Nations Conference on Science and Technology for Development*

PERGAMON POLICY STUDIES

Obstacles to Mineral Development
A Pragmatic View

John S. Carman
Edited by
Bension Varon

Pergamon Press
NEW YORK • OXFORD • TORONTO • SYDNEY • FRANKFURT • PARIS

Pergamon Press Offices:

U.S.A.	Pergamon Press Inc., Maxwell House, Fairview Park, Elmsford, New York 10523, U.S.A.
U.K.	Pergamon Press Ltd., Headington Hill Hall, Oxford OX3 0BW, England
CANADA	Pergamon of Canada, Ltd., 150 Consumers Road, Willowdale, Ontario M2J, 1P9, Canada
AUSTRALIA	Pergamon Press (Aust) Pty. Ltd., P O Box 544, Potts Point, NSW 2011, Australia
FRANCE	Pergamon Press SARL, 24 rue des Ecoles, 75240 Paris, Cedex 05, France
FEDERAL REPUBLIC OF GERMANY	Pergamon Press GmbH, 6242 Kronberg/Taunus, Pferdstrasse 1, Federal Republic of Germany

Library of Congress Cataloging in Publication Data

Carman, John Stanley.
Obstacles to mineral development.

(Pergamon policy studies)
Bibliography: p.
Includes index.
1. Underdeveloped areas—Mineral industries—Addresses, essays, lectures. 2. Mining industry and finance—Addresses, essays, lectures.
I. Varon, Bension. II. Title.
HD9506.A2C34 1979 338.2'09172'4 78-26807
ISBN 0-08-023904-8

Printed in the United States of America

Contents

Foreword

This book contains twelve essays written by John Carman between 1965 and 1978. Most of them are based on addresses delivered by proxy or in person in places as divergent as Sacramento, California; Bagauda, Nigeria; Montreal, Canada; Ankara, Turkey; and Madrid, Spain. Some are angry, even tempestuous, some gentle; some are didactic, some pleading; some are somber, even melancholic, some whimsical. All, however, are sincere and motivated by a passion for truth as Carman sees it. It is a measure of the man and his message that, unafraid to make enemies, he made so many friends everywhere.

The essays are not reproduced in chronological order. They are rather grouped as follows. The first six chapters deal with the most common obstacles to mineral development in the developing countries; the factors which dilute the value of foreign assistance, with special reference to mineral exploration; and the past as well as prospective role of the United Nations in overcoming these obstacles and, in particular, in avoiding some of the pitfalls of bilateral assistance. The next five chapters are in the nature of case studies of specific obstacles, taking a broad view of the process of mineral development, covering the gamut from the forming of perceptions of the availability of minerals to the social and economic problems of maintaining community life in mining towns after mineral resources are depleted. Within this range, these five chapters focus on the problems of taxation, which bear more than anything else on the exploitation path of mineral resources. Chapter 11 deals with small-scale mining - the mode in which mineral production started and continues in many parts of the world despite working conditions which are among the worst in any industry.

The essays appear more or less in their original form except for the deletion of references or passages relevant exclusively to the audiences for which they were originally intended. Although I have used editorial prerogative to smooth out and structure the presentation, I have meticulously avoided altering Carman's vivid style - and wit - for therein lie a joy of life, a spontaneity and an earthiness which are inseparable from the man.

The idea for this venture was born in the Delegates Lounge of the United Nations in New York, that football-field-long watering hole where Carman used to hold court at noontime more often than not. The membership of the court was flexible. The only admission requirements were partiality to Carlsberg's Elephant-brand beer, a tolerance of four-letter words, and the ability to pretend to understand what the mining fraternity - mining engineers, brokers, financial analysts, geologists, geochemists, and that class unto itself, project managers - talk about with incredible passion. After two years of working with Carman at the United Nations, I was hooked on Carmanism and, by way of perpetuating the vice, I told him: "John, I would like to edit your papers." I know he had been thinking of publishing a selection, and I feared that he would put it off forever, for Carman is a man of the battlefield, not of stuffy war-game rooms. The conversation then drifted to some mole hole in some copperbelt in some continent. When I got up to leave, John Carman walked me to the elevator and said, "You have carte blanche." I said, "Fine. Send me all your papers," and we shook on it. Ten months, fifteen letters and a three-martini encounter at the New York City Princeton Club later, the book took its present form.

What is the message of these papers and why is it important to bring it to the attention of a larger audience today? The core of the message is rather simple. In Carman's words: "It is clear that unless the current global recession continues indefinitely new mines must be found and brought into production constantly if only to meet the needs of an ever-expanding population. And, increasingly, unless we are reduced to mining almost barren rock in countries such as the United States, in the long term these mines will have to be located in developing regions... While porphyries grading as low as 0.29 percent copper are opened up in the United States, about a dozen exceptionally fine deposits running on the average about one percent lie neglected in Chile and Peru. Of course, the mining companies do send their geologists to these countries. The deposits are examined with monotonous regularity, but the investment dollars stay in the banks in New York, London and Tokyo." The reasons for this are complex. We have, he continues, on the one hand, "a climate of indifference and fear, with the major mining companies marking off most of the developing world as being unsafe for investment" and, on the other, a pervasive "suspicion on the part of governments of the motives, ethics and methods of the large mining houses, a suspicion which is more than amply reciprocated when said mining houses size up governments. There is an air of unreality to this damaging impasse."

The most insidious obstacles to mineral development in the developing countries - the ones that invite Carman's wrath - are not the high cost of mine finance or the challenge of overcoming nature in remote areas. They are, rather, what he calls "a swirling cloud of mythology, embraced largely by governments...tunnel vision on the part of those who have enough power to affect badly needed changes... . In the developing countries, which greatly need the benefits which can accrue from intelligent mineral policy, a regrettable tendency to elevate intransigence, suspicion and hypersensitivity to political considerations - real or imaginary to an exalted status. Usually this is then compounded by living in a dream world of hope divorced from reality." The obstacles, in short, are misconceptions. In the first chapter Carman presents with a searing pen and in the staccato language which decision makers understand best an overview of the most dangerous misconceptions. No group of countries or interest group has a monopoly on these erroneous beliefs and none is spared criticism by Carman. The true inhibitors of investment are misguided notions of the profitability of the mining industry and how best to tax it, of the discounted present value of minerals in the ground, of safety, sovereignty and stability, of the prerogatives and role of governments, of equating foreign assistance with neocolonialism, of the inability of developing countries to go it alone, etc.

Carman's views on these notions are anything but equivocal "...mining is not a very profitable industry. Certainly, there are many others - hamburger chains, computer software, liquor stores - which have far better records of earnings and with much less risk." "When harsh royalties are imposed," conservation often suffers as a higher grade of ore has to be extracted. In the end, the loser is the owner, in most cases the nation. ... At current rates of interest a proved mineral deposit containing billions of dollars worth of minerals has a present value of zero unless it is brought into production within ten or fifteen years...." It is the "responsibility of governments to help their citizens who are living now by intelligent exploitation of the nation's resources.... Succeeding generations will not be impoverished if the job is done well.... There is no way of assuring absolute safety of investment. There never was... Government takeover of mines will in the end push up the price of minerals because the bureaucrat is so hedged in that he cannot be a low cost producer.... No nation can ever be sufficient unto itself and, to the extent that it is not, its sovereignty will be breached. The compromise necessary to resolve and avoid conflict will always tend to appear as infringement on sovereignty."

If many of the common misconceptions are not corrected, Carman argues, all will suffer. Insistence on unrealizable conditions will push prices up and stimulate the development of otherwise uneconomic resources in countries offering less exigent terms. Once the big transnationals have secured the raw materials for their manufacturing subsidiaries, they will have a vested interest in slowing down mineral exploration and development in the Third World, for they can hardly be expected to welcome competition. Also, the relative stagnation of new mining investment in the developing countries will tend to strengthen the bargaining position of the industrial world.

Aside from a change of attitudes and reassessment of self-interest, accelerating mineral development in the Third World requires a great deal of financial and technical assistance - above all, efficient and intelligent assistance. Carman is happy with neither the volume nor the quality of such assistance. "The amount of foreign aid which has poured out," he observes, "has hardly been of sufficient magnitude to force the taxpayer to bypass the pub on his way home." Moreover, both donor and recipient have been responsible for diluting the effectiveness of the assistance. Carman is particularly critical of the common mistakes of donors, for many of the weaknesses seemingly endemic to developing countries can be traced to or have been perpetuated by bad external advice such as the "think big" approach. He is especially disturbed by the "high cost" of foreign assistance. "What greatly inflates the costs are the various allowances, subsidies, overheads and backstopping charges. One has to be shocked at such boondoggles as clothing allowances, particularly for geologists who would hardly go into the field sporting Hermes ties and Brooks Brothers suits. That such allowances do exist implies that in the home country clothes are not de rigueur."

The positive record of the United Nations in the field of minerals, especially exploration, is little known or appreciated outside the mining fraternity. Carman documents that record authoritatively, for he has helped to build it himself over two decades. He argues strongly that the role of the United Nations, including the World Bank, should be strengthened and expanded. The United Nations Carman talks about is not that of endless speeches and resolutions, it is that of technical cooperation; it is not that of demagoguery, it is that of multinational experts united to solve technical problems whether they involve the drilling of an exploration hole in difficult terrain, the optimum processing of a mineral ore, or the drawing up of an appropriate mining legislation.

Despite the facade of fire and fury, Carman's approach is essentially pragmatic. His pragmaticism does not mean accepting the status quo philosophically. It means, rather, taking a broad and long-term view of one's self-interest. To illustrate, he would argue: If you, the host government, overtax the mining industry, be aware that it will skim the cream off the deposits and will not invest in future production. If you, the company, manage to exact stringent terms from the host government, you should not realistically expect to operate in that environment for a long time. Carman is not against government intervention; he is against stupid government intervention. He is not against private industry when it shows vision and statesmanship. It is a sad commentary on the fashionableness of the advocacy politics of today that it takes more courage to be neutral or even-handed than to take sides. Those who are, unfortunately, do not have lobbies.

Carman's values emerge clearly in these chapters. First, there is a respect for expertise - for a job well done - regardless of the national origin, creed or color of the expert. He attacks xenophobia and paternalism toward the cadres of the Third World equally. In fact, he has high respect for the talented people of the developing countries, because he is aware of the enormously difficult conditions under which such talent is born and labors. Second, he has a pride of profession and a genuine love for the United Nations where he has worked for nearly 20 years. "I would not change a day of those years," he says, "I have had a ball." Finally, although he has dealt with material things and cost-benefit analysis all of his professional life, he has never surrendered to a professionally narrow view. In 1972 (Chapter 7) he wrote:

> I firmly believe it will not be lack of mineral resources which will put an end to the two cars in every garage dream, to development of outboard motors with the power of a Mack truck, to garbage cans filled to groaning each day with container debris. It will be the fumes in the air and the slick on the sea which will finally force a halt if there is to be meaningful survival. The high tide of sheer materialism may be very close. Then, as the act of possessing abates, a fundamental truth may dawn upon human consciousness: once basic needs for nourishment, physical, mental and aesthetic, are met and shelter is assured, the quality of life will be found to be only superficially related to goods and chattels.

The power of the essays comes from a sharp eye, empathy and wide field experience. Whatever the area he writes on, you know he has been there, and you know where his heart is when he says, for example, in Chapter 11, concerning small-scale mining:

> It is also a brutalizing business. Labor is almost invariably exploited. Submarginal wages are paid, and working conditions are too frequently appalling. I will never forget the tortured faces of miners in the Andes, with 50 kilos of rock on their backs struggling up steep inclines with vertical rises of as much as 200 meters, and the families of workers living in the mouths of abandoned adits. Goats were better housed and, in terms of value, probably rated higher.

And he shares his experience with compelling immediacy when he recollects:

> The shaft was vertical and over 200 meters deep. The "cage" consisted of a piece of hard rubber about the size and shape of a bicycle tire which was connected to a small, 1890s vintage hoist by a quarter-inch steel wire rope red with rust.... Not surprisingly, the rope broke about two weeks later, dropping two men to their deaths. Another "cage" which is often encountered consists simply of an oil drum. The mere act of getting in and out of such a contraption has decidedly dangerous elements. In effect, too often life is cheaper than equipment.

His essays speak forcefully to the large audience of thoughtful people engaged in all phases of resource development. But they hold a special message, it seems to me, for those who are about to embark on a career in mining. His message to them is: Your field will be as exciting as your spirit will allow. Make the world your office and enjoy what you are doing. But above all, do it with an open mind and an open heart. Then you will do it well.

- Bension Varon

Acknowledgments

The journey toward this book started many years ago. During World War II, I spent almost five years in the engineer corps of the Canadian Army, concerned on occasion with such activities as training men to handle Bren guns and throw grenades, pushing Bailey bridges across gaps created by enemy demolition squads, laying mines, lifting them - interesting work but not in anyway relevant to opening holes in the ground to extract minerals. Consequently, when the war was over I undertook some post-graduate work at Columbia University to get back on track. There I met Elmer Pehrson, then in charge of foreign activities of the United States Bureau of Mines and a visiting lecturer at Columbia, giving a course on mineral economics on Saturdays. Through him, I met the late Gilbert (Slim) Monture, also a pioneer mineral economist, who often acted as Canada's official representative at international meetings dealing with minerals. Both men, brilliant and exciting, had an enormous and lasting influence on me. They made me see mining as something much more than a purely technical process for finding and extracting minerals. They instilled in me some understanding of the complex interactions involved in meeting mineral needs - a play of forces in which the physical aspect depends upon what goes on in the minds of men who put up the money, provide the brains and muscle and determine the character of the political and economic climates in which mineral resources are developed and put to use. I will be eternally grateful that these two men taught me to look at the whole context and not to view a mine in isolation.

When the time came a quarter century later to put between covers what the journey had yielded, I had to rely on the help of a new generation. First of all, because of the slashing blue pencil wielded by the editor, I had to call on the assistance of several more than competent secretaries who at one time or another had worked with me. They did heroic work in making splendid copy from scribblings that frequently were almost indecipherable. While it is unfair to single out any particular person, I must give special thanks to Alice Coquia who, with her physician husband, represents the enrichment conferred upon American society by people from the Philippines. Gratitude also has to be expressed to Pavel Kral of the United Nations Centre for Natural Resources, Energy and Transport for drawing the charts. In addition, a great deal is owed to Wolfgang Gluschke, also of the Centre, for being a relentless critic with the annoying habit of almost always being right. He supported this undertaking from beginning to end and was instrumental in getting Pergamon Press interested in publishing the book.

It must be confessed that the whole exercise has not been devoid of actual embarassment. I claim noble lineage as the cultural descendent of Milton, Shakespeare, Byron, Dickens et al. It has therefore been somewhat humiliating to have the editor, a Turk, teach me a great deal about how to use my mother tongue properly.

Finally, Elizabeth Carman has been stellar. As a critic of the silent school she is a nonpareil; instead of verbal torrents, the raised eyebrow studied pause, questioning stare. These have effectively, although not totally, dampened my addiction to hyperbole and the sweeping statement. For years she has been beset by the cacophony of an enthusiastic one-finger typist and bedevilled by his propensity to spray papers to all four winds when settling down to write. Such vexations increased geometrically as all was drawn together to provide Pergamon Press with a manuscript. Throughout she accepted the confusion and annoyance with stoicism and grace.

Obstacles to
Mineral Development
(Pergamon Policy Studies-18)

1 Obstacles to Mining Investment in the Developing Countries*

Much is heard these days of the confrontation between the governments of developing countries and international mining capital. Sadly, there are, in actual fact, few cases of anything as positive as confrontation because such an action at least signifies a dialogue which may begin badly but end with positive results. Instead, we have a climate of indifference and fear, with the major mining companies marking off most of the developing world as being unsafe for investment. There is an air of unreality to this damaging impasse. The common sense and objectivity which are prerequisites for positive action are lost in a swirling cloud of mythology, embraced largely by governments, which inhibits mining investment in the developing countries. Let us examine some of these dangerous misconceptions.

"FAT CAT" PROFITS

Contrary to nearly universal opinion, when viewed over a period of years, mining is clearly revealed as not being a very profitable industry. Certainly there are many others - hamburger chains, computer software, liquor stores - which have far better records of earning and with much less risk. To illustrate, a number of years ago I calculated the return after taxes of the entire Canadian mining industry for a year

---

* Based on a paper on the same subject prepared for the Workshop on New Directions in Mineral Development Policies organized in Bagauda, Nigeria, by the Association of Geoscientists for International Development, September 8 - 10, 1975, and presented, with some changes, at the Annual Meeting of the American Institute of Mining and Metallurgical Engineers in Atlanta, Georgia, March 6 - 10, 1977.

which was neither booming nor marked by recession. Per dollar of sales, earnings were slightly less than ten cents. This was the average for the successful enterprises. If, from that modest profit, one were to deduct the losses of the firms which had no profits against which unsuccessful exploration and mine development programs could be charged, the average return to the industry would not exceed the interest which could be gained by investing in sound securities.

What confuses the layman are the bonanza years. Occasionally, there are periods of very high profits. Almost invariably they are viewed by governments as being exorbitant. They are, however, simply a function of the volatility of minerals markets. If, over a period of a few months, a shoe factory were able to treble prices while maintaining volume, its profits would also appear to be extreme. But with minerals the price usually falls as drastically as it rises. In terms of constant dollars, the average sales price of metals today is probably slightly lower than it was at the beginning of this century. It is strange that in the depressed years when profits are at best modest, governments do not express concern even remotely proportional to the outcry of indignation which is heard during periods of high earnings.

THE ROYALTY SQUEEZE

With the state being viewed as the owner of subsoil resources almost everywhere, there has been an alarming increase in the imposition of royalties based on the value of production and, in some cases, payable in the form of the mineral itself. Royalties are viewed by governments as a fair and rightly deserved tax. They are quite the reverse, however, because where they simply constitute a direct operating cost, conservation often suffers as a higher grade of ore has to be extracted. In the end, the loser is the owner, in most cases the nation. Particularly pernicious is the imposition of royalties whether a profit is being realized or not, as opposed to the treatment given to other industries. If the trend to impose ever-higher royalties continues, it is not difficult to visualize the closure of a large number of mines or nationalization by enforced bankruptcy. I suspect that more than one government has the latter course of action in mind. A good example of what is involved is a statement in the "Report for the Six Months Ended March 31, 1975" of Teck Corporation, a medium-sized Canadian mining company:

> It is interesting to note that Beaverdell operated at a loss of $57,000 during the second quarter, but it nevertheless incurred a tax liability of $25,000 under the current British Columbia royalty and mineral land tax situation.
>
> Provincial royalties payable on oil and gas production amounted to $1,800,000 during the half year. It is estimated that the Federal Government collected an additional $2,000,000 as a result of its Export Tax. Since Ottawa has refused to recognize the Provincial

royalties as a cost, the producer is subject to Federal Income Tax on the amount paid by the company to the Provinces. It is called "income", but it sure doesn't feel like it! Under the current situation, an increase of $1.00 per barrel in the prices of Saskatchewan oil would cost the company $1.10 in royalties and taxes.

The Canadian price of oil was increased in the Spring of 1974, ostensibly to encourage exploration in the interest of increased supply. Unfortunately, as was pointed out at the recent First Ministers' Conference, more than 95% of the increased revenue was skimmed off by various Governments. "Too many hands reached into the pot", and 95% of the increase was wasted - at least with respect to the purpose for which it was intended.

Ironically, until a few years ago, Canada was viewed as about the most favorable country in the world for mining investment. A few years from now it will be interesting and quite probably horrifying to observe to what state public exactions have reduced the industry.

THE "THEY HAVE TO COME TO US" SYNDROME

In too many places, the existence of superior ore bodies has been viewed as an irresistible lure which, in the end, will force the international mining houses to invest. In the end, yes, but if that means 25 years from now, a lot of people may die of malnutrition in many places before any value is obtained from the nation's mineral resources. A few years ago British Columbia, Canada, was considered as a "safe" environment for investment. After perhaps a billion dollars of investment in porphyry copper deposits became locked in, it became painfully evident that it is anything but safe because of the royalties mentioned above. After imposition of the new tax regime in 1974, development spending and exploration drilling in the mining industry of British Columbia dropped by about 99 percent - literally. Capital demonstrated clear intent to have nothing more to do with a hostile province. It would hold on as long as possible, but it could hardly be faulted for the refusal to risk exposure to more punishment.

While porphyries grading as low as 0.29 percent copper are opened up in the United States, about a dozen exceptionally fine deposits running on the average about one percent lie neglected in Chile and Peru. Of course, the mining companies do send their geologists to these countries. The deposits are examined with monotonous regularity, but the investment dollars stay in the banks in New York, London and Tokyo.

THE NONRENEWABLE MYSTIQUE

In general, minerals do not replenish themselves in place or in time spans which have any meaning. Some, such as hydrocarbons, suffer total dissipation when used. Others are amenable to recycling to some degree. But, on the whole, mineral deposits are rightly viewed as depleting assets. This arouses a great deal of emotion. Witness the wailing of the Cassandras which arose with publication of The Limits to Growth (Meadows et al., 1972). It is now recognized that "...the book... has an air of spurious scholarship throughout. It could be mistaken for elaborate satire and is pseudo-science at its worst" (London Times, March 31, 1973). Nevertheless, there is widespread fear that man will eventually exhaust his mineral supplies. This, incidentally, will take a great deal of doing if only because the world is entirely mineral. D.B. Brooks expressed very clearly where the danger lies: "Contrary to conservationist tradition, I see most non-renewable resources as all but continually available over time, whereas the supposedly renewable resources such as air and water are in jeopardy of depletion" (1973).

But let us be pessimistic by envisioning a day three or four hundred years hence when the world could be faced with a shortage of copper, lead, nickel, what have you, which can only be overcome by production from rock so low in grade that the price has to be exorbitant. Should we sacrifice this generation and the next because of hypothetical hardship which just might possibly affect the fifteenth generation from now? Does it make sense to jealously guard mineral resources, to put them in the refrigerator, so to speak, under the guise of protecting the national patrimony? Not at all. At current rates of interest in most countries, a proved mineral deposit containing billions of dollars worth of minerals has a present value of zero unless it is brought into production within ten to fifteen years. It may be an enormous mass of valuable minerals, but if it is to lie fallow for two or three decades it is not a national asset. It can be defined as "potential" only something which does not feed and clothe hungry people. What I wish to emphasize is that the responsibility of governments is to help their citizens who are living now by intelligent exploitation of the nation's resources - land, water, mineral. Succeeding generations will not be impoverished if the job is done well.

THE "SAFE" COUNTRIES

During the past ten to fifteen years, international mining capital has concentrated upon a handful of countries because these appeared as politically stable places with full respect for sound business practices and the sanctity of contracts. Lately, finding such countries has been very difficult. The last three or four years in particular have been traumatic for the major mining houses.

Ireland experienced a mining boom in recent years which was

extraordinary for a country which had produced nothing but building stone and peat during this century. The prime mover was a 20-year tax holiday. Unfortunately, the mining companies found themselves beset rather abruptly. After only a few years of profitable operations, they saw every discounted cash flow calculation thrown out of the window by the arbitrary imposition of taxes on profits about as onerous as anywhere else in the world. I have already referred to the parallel events in Canada. Australia enjoyed a mining boom for about ten years. The advent of a new government in 1974 seriously changed the investment climate in a matter of months although confidence has been partially restored with the repudiation of that government. At the moment, the mining community views only two of the major mining countries as attractive for investment: the United States and South Africa. The former may not long warrant this categorization because of the high cost of meeting environmental standards and the latter seems headed for a stage of siege.

What the foregoing adds up to is that mining capital, which has always operated in a high-risk environment, will be faced with new challenges in the years ahead. It will no longer suffice to seek out compliant and moderate governments which offer to more or less guarantee the protection of the investment from sociological, psychological and political waves. With the long lead time required today for a significant mining investment, the company can confidently expect that the contract negotiated with a government will be reviewed three or four times before the mine produces a pound of saleable product - if it is permitted to persevere to that point. Consequently, the mining company of the 1970s should bear in mind the following:

1. There is no way of assuring absolute safety. There never was.
2. The game is still worth playing because the world has to have minerals, and the more the timorous turn to less risky pursuits, the greater the potential for interesting profit opportunities, provided that governments do not pounce immediately. In addition, there is a fast-increasing tendency for governments to take over operating control of mines. In the end this will certainly push up the price of minerals because the bureaucrat is so hedged in that he cannot be a low-cost producer.
3. There is, however, the "safety in numbers" gambit. Instead of having one or two massive ventures, placing a limited number of eggs in a lot of baskets warrants consideration. The trend toward consortia, toward financing by a score or more of bankers and having operations in a dozen countries, is a reflection of appreciation that the "shotgun" approach cannot guarantee the health of any particular mine but can go a long way toward protecting the viability of the total operation.
4. Eventually, if only because of rising demand, linked to superior grades of ore and unexplored potential, accommodations must be made whereby investment in the mineral deposits of the developing world will have to rise relative to that in the industrial countries.

NOSTALGIA FOR THE GOOD OLD DAYS

Not much needs to be said under this heading except that it would be advisable for mining companies to realize that the good old days will not come back again. In isolated instances - Bougainville, Papua New Guinea, being a good example - a true bonanza situation may be encountered, but it will not endure for long. Increasingly in the future, overseas investment will be pursued more in order to supply the semi-fabricating arms of the corporation than for the sake of profit.

THE NEOCOLONIALISM HYSTERIA

By and large, the more newly independent a country, the greater the likelihood of suspicion of the motives of the foreigner who offers to invest. The potential investor will almost invariably be interested only in the profit potential, but the host nation will fear that he is trying to obtain by economics the mastery that he enjoyed in another day by force of arms.

"Neocolonialism," "infringement of sovereignty," "protection of the national patrimony" - these become catchwords. In my view, they are very misleading, dangerous and demagogic. No nation can ever be sufficient unto itself and, to the degree that it is not, its sovereignty will be breached. The foreign enterprise coming to a country will expect rights that are based on international law and practice which frequently will not be in harmony with national laws. Resolving the contradictions will almost always necessitate compromise which may appear to constitute infringement of sovereignty. This is particularly so with the mining industry. Throughout its history, it has been developed initially by foreign capital. In the nineteenth century, it was largely British money which put American mining on the road to greatness. To this day, capital from the United States owns better than half of Canadian mining although control is vested in Canadian legislative bodies. The same pattern exists and will prevail in developing countries until the know-how and capital can be built up to permit national ownership and operation of mining enterprises. It is ironic indeed that in a chauvinistic world, at least in the beginning, much of such know-how will have to be contracted from abroad and financing arranged by borrowing from foreigners. Consider the efforts by the Soviet Union to enlist the participation of private foreign enterprises in developing its natural resources.

THE "WE CAN DO IT BETTER" DELUSION

Where nationalism is particularly virulent, the degree of self-delusion at policy-making levels can be appalling. To illustrate, a country (which shall remain nameless because the legislation I shall refer to is still in the drafting stage), in writing what it considers most erroneously to be modern mining legislation, has included a clause which specifies that within five years of the start of operations, 100 percent of the personnel will have to be nationals. And this in a country where the principal public organization concerned with the industry, the mines department, does not have a mining engineer or a geologist on its staff! In addition, there has been absolutely no opportunity for the very few national earth scientists to obtain any managerial know-how in regard to extractive operations. That such a clause could be considered is in itself very revealing. It shows extreme ignorance of the principles and practices which guide middle and top management. Also, it slights one of the basic policies which has made certain nations industrial powers: maintaining a constant search in all parts of the world for special skills. Any vibrant society needs cross-fertilization. No nation is self-sufficient in regard to talent. The welcome extended to the likes of Einstein, Steinmetz and von Braun in the United States emphasizes the point.

THE "SAVING FOREIGN EXCHANGE" PITFALL

Frequently governments in the developing world bring into production mines which stand not the slightest chance of being economic, either because of grade, size or location or a combination of all three or more. The rationale that the project is worthwhile because it will result in saving foreign exchange is far too often a snare and a delusion. In most developing countries with an operation of any real importance, the bulk if not all of development capital will have to come from abroad; managerial skills and technical expertise, sometimes down to the level of skilled workers, will have to be imported, usually with a fancy price tag, and the vast majority of operating supplies and equipment will not be produced locally. At best, such operations may have a marginal impact upon easing balance-of-payment problems. More often, there will be a net outflow of foreign exchange, at least during the years when the foreign loans have to be serviced. Inevitably, a self-defeating mechanism starts up. In any tight economic situation, corners to be cut are avidly sought. In these instances, the foreign staff is pared long before the national replacements are up to the job, the plant and equipment are allowed to run down because the government simply does not have the funds to replace worn out or obsolescent equipment. Indeed, a common characteristic is the inability to maintain even minimum levels of spare parts and operating supplies. As inefficiency waxes, the mine becomes ever more high cost and often stays in operation largely at the expense of the worker who is paid only subsistence wages, if that, who does not have the tools to work with, and who cannot therefore enjoy the incentives and spiritual satisfaction which come from association with well-managed and viable enterprises.

OTHER INHIBITORS OF INVESTMENT

1. The magnitude of investment required to bring a major mine into production becomes ever more staggering - in too many instances, billions rather than millions. Years of negotiations with potential lenders, generally the internationally oriented private banks, may be needed before commencement of the development of a known ore body. Then, too often during the construction phase, it becomes apparent that the generous provisions for inflation and contingency simply will not suffice. So, once again the struggle is joined to successfully arrange for refinancing.

2. A vital problem revolves about the suspicion on the part of governments of the motives, ethics and methods of the large international mining houses, a suspicion which, incidentally, is more than amply reciprocated when said houses size up governments. Neither party lacks justification.

3. Legal complexities very often constitute a morass in which investment opportunities founder. In many places it is difficult to determine what the law is. There is often a tendency for successive acts to overlap and contradict without superseding. Too frequently the administrative regulations required to give life to a statute are not enacted.

4. A corollary of the rising nationalism so characteristic of the twentieth century is the intense desire in the Third World to industrialize at all costs. Thus, negotiations may occasionally become stalemated because of the insistence of governments upon processing of minerals to a point well beyond that which makes sound economic sense.

5. In recent years there has been an alarming increase in the self-destructive trend on the part of governments toward trying to negotiate exploration contracts which do not bear concomitant rights to exploit. Any serious mining company is painfully aware that such an agreement will place the company at the mercy of the government if an ore body is found; and the richer the ore body, the more exigent the conditions are bound to be. A classic illustration of what is involved here is the Cerro Colorado case in Panama, one which has not gone unnoticed in other parts of the world. After completion of the exploration phase, negotiations regarding an exploitation agreement dragged on. Finally, the Panamanian government took over the property with the announced intention of indemnifying the company at least for the total amount spent on exploration. This could hardly be considered as satisfactory by the company, because the recoverable value of the copper in the deposit approaches ten billion dollars even at today's (early 1977) depressed metals prices.

6. The developing countries are overoptimistic when assessing their bargaining position. Unless locked in by earlier investment, the mining company cannot be bludgeoned into pouring in money. Kennecott Copper, after spending fifteen million dollars in Papua New Guinea to find and explore the Ok Tedi porphyry copper deposit, recently withdrew from that project renouncing all rights. The deal foundered because the government of Papua New Guinea would not agree to international arbitration of disputes, and the taxation rates were simply too high for the risks involved. Kennecott had agreed that the government should receive the lion's share of any windfall profits but required a reduction of the tax base for normal profits. This was not accepted; hence Kennecott's departure. In fact, more and more mining capital is being invested to achieve horizontal integration designed to offset the high-risk nature of mining which, too often, is exacerbated by the failure of mineral prices to keep abreast of long-term inflationary movements. Hence, a great deal of mining money is being invested in real estate, timber reserves - anything which offers reasonable security and a relatively high return.

7. Finally, momentum has decided weight. As I noted earlier, for many years mining investment has been concentrated in relatively few countries. Outside of the centrally planned economics and excluding stratiform deposits such as coal and limestone, two-thirds of the world's major mines are found in four countries - the United States, South Africa, Canada and Australia. And while the two latter nations have been undergoing reappraisal of mining policy in recent years - usually in ways patently unfavorable to the mining industry - investment proceeds at a far higher rate than in developing countries if only because of the pressures stemming from an abundance of know-how and infrastructure and the existence of a very large mining industry which usually forces a considerable amount of reinvestment.

CONCLUSIONS

I strongly believe that the relative stagnation of new mining investment in the developing countries tends to improve the bargaining position of the industrial world. With the exception of petroleum producers, very few Third World countries are able to build a strong enough financial base to negotiate on reasonably even terms with the importing nations. The insistence upon unrealizable conditions may safeguard mineral reserves by keeping them fallow but often only at the price of stimulating the development of otherwise uneconomic resources in countries offering less exigent terms and greater stability.

It should not be overlooked that the principal mining companies, many of which are fully integrated vertically, will have a vested interest in a slowdown of mineral exploration and development in the

Third World once their mineral requirements are secured for the foreseeable future. In theory, the international mining community should press for rational global actions which are aimed at improving the supply-demand situation. In practice, the strong mining concerns with assured supplies to satisfy the needs of their fabricating subsidiaries or relatively secure long-term sales contracts can hardly be expected to welcome new mining operations, either at home or abroad, which can make a sizable contribution on the supply side.

The final force which sets a limit to the share of profits and control which governments may obtain in the initial capital-recovery years is the banker. In earlier days, mine financing came largely from equity. Today, the usual vehicle is a consortium of banks which may total scores and be chartered in a dozen different countries. So, in contrast to the often reckless and occasionally even fatal approach of a simpler age which was exemplified by "let's take a chance," today's significant mines come into production only after the bankers have dispassionately, thoroughly, and, by and large, pessimistically analyzed all elements which determine discounted cash flow. If returns are seen to be relatively modest after all the safety factors have been applied in respect to anticipated risks, no investment will take place.

What emerges is a generally negative picture from the viewpoint of the developing countries, and with very good reason, because the world's mining industry is replete with anomalies, waste, inefficiency and above all tunnel vision on the part of those who have enough power to effect badly needed changes. In the developing countries, which greatly need the benefits which would accrue from intelligent mineral policy, there is a highly regrettable tendency to elevate intransigence, suspicion and hypersensitivity to political considerations, real or imaginary, to an exalted status. Usually this is then compounded by living in a dream world of hope divorced from reality. In the end, such attitudes redound only by increasing the profits of mineral producers in the well-mineralized industrial lands. On the other hand, countries such as the United States and Canada, by depending upon what would normally be uneconomic mineral deposits if the world achieved something even remotely approaching a global strategy for the provision of mineral needs, are certainly perpetuating the "feast or famine" pattern so long characteristic of minerals availability.

2 Dilution of the Value of Foreign Aid With Particular Reference to Mineral Exploration *

All human endeavor has its built-in and somewhat characteristic inefficiencies. The cost of building a skyscraper is calculated down to pennies and then a flat 15 percent is added on for contingencies. This, plus the later overruns which are difficult to explain, constitute the votive offering to inefficiency.

In this chapter, I shall examine the more prominent weaknesses which dilute the value of contributions to the development of less fortunate parts of the world, limiting the discussion to aid from the public sector, bilateral or multilateral. Both recipient and donor are quite capable of reducing effectiveness, and I shall consider each separately.

THE RECIPIENT

Manpower

The recipient's participation in programs or projects supported by foreign aid can vary from peripheral activities, such as clearing items through customs or fueling aircraft, to the other extreme where national counterparts will outnumber the foreigners by twenty to one and be deeply engaged in every facet of the project. Here is a major cause of diluted effectiveness. Too often their drivers tear vehicles to pieces; mineral analysts get absentminded; geologists bring into play a

---

* Based on a paper entitled "Limiting Factors in Foreign Development Aid with Particular Reference to Mineral Exploration", published in the Proceedings of the Twenty-fourth International Geological Congress held in Montreal in 1972.

multitude of ploys to avoid going to the field, misinterpret data and do work which is so substandard as to be worse than valueless. Of course, such sinning is not unheard of in industrial nations. Nevertheless, the incidence tends to be higher in those places where lack of physical resources, sustained programs and continuity of leadership inhibit the growth of a sound philosophy of work.

It is very easy to criticize but one should have the good sense to appreciate the background. For example, the professional man in many developing countries too often has never really had a chance. A common pattern is graduation from an undernourished university and employment by a public organization with funds so limited that most personnel become office-bound, there simply not being sufficient resources for sustained field programs; and too many people cannot go to the field in any case because they have to have two or even three jobs to make ends meet. Those who are fortunate enough to be able to learn their trade in the time-honored way usually face the handicap of having to operate with outmoded and insufficient equipment, if, indeed, certain vital elements exist at all. In many, if not most, of the geological survey organizations in developing countries, the diamond drill or any sophisticated geophysical instrument was unknown until very recently. Back-up tends to be inadequate due to lack of well-stocked libraries, petrological labs, drafting and printing facilities, or even moderately efficient analytical laboratories.

Then there are the psychological considerations, the best illustrations of which can be found in countries with a Latin culture. In general, the prestige professions are law, medicine, arts and letters. To get established in these, one is well advised to have connections. Accordingly, the young man who sweats his way through mining engineering or geology frequently comes from a lower social stratum. As insecurity is the breeding ground of snobbism, a situation is created whereby clean hands are a status symbol. So, the job of slogging up the ladder as field geologist or shift boss is not considered attractive. Too often the phenomenon is diagnosed as laziness. It is quite the reverse. The Latino is an extremely energetic person. The average citizen from the industrialized part of the world would collapse with battle fatigue if he followed "Juan Gomez" through a quiet day: mad dash to check in at the ministry on time, whirl senseless papers around for a while, lunch with somebody who just might ask him to examine his property on the weekend, off to the university to give his class, and then spend the evening translating business letters for a client. There is, of course, some element of exaggeration here, but not a great deal.

Where a caste system exists either implicitly or openly, mounting a safari can be something so tremendous that even Cecil B. DeMille would have recoiled. As the numbers required to do even a routine job increase, efficiency tends to sink in geometric proportions.

Finally there is the "one-eyed man in the kingdom of the blind." Postwar arrival of independence found many countries with only a handful of university graduates - and the overwhelming majority of these did not have science degrees. Let us consider a hypothetical country where there were only one or two geologists. Almost by

Ill. 2.1. Diamond drilling in Tamil Nadu an example of excessive employment of labor. Credit: United Nations / T.S. Satyan.)

definition these people would be quite young, probably in their twenties. In wealthier environments, they would simply be bodies to be sent into the field to learn their trade, representing initially a financial loss to the employer. Instead, in a land in a sad state of educational preparation, any graduate professional is such a rare being that he tends to move quickly to a rank which is far above his true capabilities. This state of affairs is then compounded by the insidious intrusion and adoption of one, more, or all of Parkinson's Laws.

Turning to the skilled worker, let me take the driver as illustration of the weakness of this group. He too often makes the error, frequently disastrous, of confusing manipulative ability with proficiency. That he can corner like Stirling Moss is all that matters. If in the process tires are shredded, transmission debilitated and springs broken - "what the hell," "c'est la vie," "que cosa," a shrug of shoulders. In many cases he is a surprisingly ingenious mechanic, but too often such facility is a product of his own lack of even elementary foresight. Again, understanding is required. If you have never personally paid a repair bill, the internal combustion engine is a passport to greatness, a thing of glory. If something goes wrong, you get overtime and the company pays.

Finally there is the laborer. Here there is seldom cause for complaint. Given reasonable nourishment, such workers could often be top hands in industrial countries, union codes permitting. Unfortunately, in an increasingly technical world, these people do not constitute a "make or break" component in most projects.

TRAINING

In recent years, great and too often misguided effort has been exerted to upgrade manpower in the developing countries. For example, virtually all United Nations projects allocate funds for fellowships. To be quite blunt, it is more than frequently a waste of money because:

1. The fellowships are given to the wrong people - the professionals. In general, such people know the theories; what they need is practical experience.
2. Far too high a proportion of fellowships are for graduate work at universities. If the "clean hands syndrome" is a problem with the bachelor of science from a national university, a graduate degree abroad only reinforces this psychological block.
3. Diplomas are often so cheap as to represent little. Even good schools may relax standards for students from developing countries when there is a linguistic problem. "He did not understand the question" is a frequent excuse for wrong answers.
4. At the other extreme, when the degree is good and has been well earned, the fellow too often will not return home.

5. The most impoverished countries simply do not have qualified candidates for fellowships. One large African country can be cited which graduated ten students in all from secondary school in 1968. It is thus not surprising that of recent years the countries which have received the largest number of United Nations fellowships have been comparatively developed. The rich get richer.

6. There is far too little on-the-job training. The efforts of the Canadian government and the United States to arrange training tours for fellows are most laudable. However, far more functional in my opinion is the type of assistance which mining companies have provided to the United Nations. They have taken on trainees and attached them for months to field parties, paying the going rate for geologists of equivalent rank. Of course, such trainees are not expected to be significant contributors, but when they finish their stint, they are immeasurably more valuable to their own country.

7. Finally, I propose that often the best candidate for training abroad is the subprofessional. On one United Nations project which involved regional geochemical reconnaissance of about 140,000 square kilometers, approximately 50 field assistants were employed at the peak. These were young men with some years of secondary schooling and even, occasionally, with one or two years at university. After learning to read aerial photographs in the sense of a locational map, they did all the field work of collecting the stream sediment and soil samples, and they did the job well. By the time the program ended, most were skilled and enthusiastic prospectors; a handful could be rated as geologists. With such people, a practical training course abroad could in many instances tip the balance, making them more than functional lifetime members of the fraternity which seeks mines.

Finance

With almost all aid programs, recipient governments have to assume financial obligations. In the ideal instance, e.g., most former British colonies and some Francophone countries, the system dictates that no obligation can be incurred unless incorporated into the annual budget. If a request arrives from such places, the donor can be sure that the required counterpart funds have been allocated. Unfortunately, there are other countries where no such safeguards exist. A ministry will enthusiastically seek funds for a project. After aid is approved by the donor, it will be found that no budgetary provision has been made for the counterpart contribution. Affairs may be stalemated for a year before the required contribution is deposited in the bank.

Government support is provided in two forms: counterpart in cash and counterpart in kind. The latter, in effect, amounts to using already existing resources for the purposes of the project. Space is made available for the internationally recruited team members; geologists, clerks, drivers, etc., are seconded to the project. However, the in-kind assistance can be somewhat on the mythical side. Governments often promise far more than they can honor, especially when requesting aid. Therefore, it is advisable to seek the maximum possible amount of in-cash support. Unfortunately, this is usually a scarce commodity.

All of this adds up to this generalization: with almost all projects, operations are hampered by the inability of governments to meet their counterpart obligations in time, quantity and quality. A great deal of improvisation is required, as well as frequent acceptance of relatively low standards or efficiency. In the early stages of one project, I can remember the ridiculous picture of a field party setting off into the mountains in a 15-ton truck, instead of the more suitable Land Rover.

Legal Framework

Many developing countries have a mining code which does not reflect the realities of today. Some have been basically unchanged in this century, similarly with fiscal regulations. Thus, when something of interest is encountered, years can drag by before an agreement can be reached between the government and the potential investor. The law is simply not geared to the efficient and rapid action which is so imperative to meet the pressing needs of many nations.

With old-fashioned codes designed for high-grade, small-tonnage operations, only by hammering out special agreements can an environment be created for a mass-mining project. However, it is extremely difficult to do so because many governments simply cannot put together a skilled and realistic negotiating team.

The "Top Secret" Complex

Lack of access to existing data can seriously hamper operations. Usually, the secrecy mania is most highly developed in military circles. To illustrate, in a country which will remain anonymous, maps being imported by a British firm were seized by the military, stamped "top secret," and hidden away. That the maps were on sale from the Queen's Printer in the United Kingdom for a nominal sum was deemed to be quite irrelevant. In another instance, a contract for airborne work signed by a ministry was held up for months after the arrival of the airplane because the armed forces custodian of aerial photography refused to release any material covering areas within 50 miles of an international frontier across which no forays had occurred for centuries.

However, the soldier or airman is not the only culprit. Ministries and even departments therein jealously guard documentation which should be of national and even international interest. Data become compartmentalized and hidden to the degree that their very existence is known to only a few. Consequently, the potential for senseless reworking and restudying is high.

Finally, in too many places sheer lethargy, ignorance and lack of system create irreparable erosion. The 20 copies of the report dwindle away to zero. I recall a mission to a country where no aerial photography was known to exist for a very sizable area. Having been assured by responsible people that the coverage had been completed and dispatched some years earlier, a search was commenced. A happy ending was achieved when the packing case containing the material was found in the pantry of a ministry, serving as a table for preparing the torrents of tea and coffee which flowed through the offices.

THE DONOR

Perhaps too frequently the villain is the nation supposedly spreading largesse. The gap between expressed intentions and actual performance can be formidable. Let us look at the various ways by which the value of aid is diluted.

Direct Transfer of Technology

Too often "what works at home" is applied willy-nilly without any real examination of the needs, character and peculiarities of the recipient country. To illustrate, in the industrialized world the large-scale geological map has usually provided the take-off for the detailed exploration designed to find mines. However, in many of the poorer nations of the world such maps constitute luxury articles which simply cannot be afforded. There is neither the time nor the resources for such laborious and expensive unfolding of the geological picture. Accelerators have to be sought which may not be as technically satisfactory but can develop enough data to attract capital: for example, photogeological maps, regional geochemical programs, and reconnaissance airborne geophysical surveys. Under these circumstances, one cannot feel that today's urgent problems are being brought measurably closer to solution by most mapping work undertaken with bilateral assistance.

Similarly, the "think big" approach may, in the end, result in no service whatsoever to the recipient country. Making airborne geophysical measurements over very large areas - hundreds of thousands of square kilometers, for example - can be functional in lands where there is no lack of interested capital. To do so - and it has been done - over tropical rain forest will probably saddle the government with the impossible task of completing effective follow-up on the ground. Under such circumstances, there is the very real possibility that nothing concrete will be done in the near future with the data, and in the end much of it may be lost. The problem here can be particularly serious when the donor provides aid under a soft loan rather than as a "no strings attached" contribution.

The airborne magnetometer can be used to illustrate another

common weakness in the direct transfer of technology. A rich country can easily afford close line spacing, let us say 500 meters, but few developing countries can. On United Nations surveys, among others, it has been found that stretching out the line spacing does not result in a proportional reduction of effectiveness. It is general policy to use a separation of one kilometer on the basis that there are few places in the world today where it is economically justified to seek magnetite deposits. Hence, the technique is essentially used to obtain understanding of major geological features. These will present just as clear a signature with a 1,000-meter spacing as with 500-meter spacing. Indeed, it can be argued that the kilometer separation of flight lines is unnecessarily close.

Recently, in a Canadian newspaper of a city located well within the pre-Cambrian Shield, there was a picture of members of a United Nations team panning stream sediments in Ethiopia. The caption underneath referred rather superciliously to such "primitive" methods and put in a plug for mechanized surface prospecting such as that practiced in Canada. This reflects a mentality which on more than one occasion has been inflicted upon developing countries by well-meaning but custom-bound foreign technicians. In the first place, the pan is a very useful exploration tool for a variety of heavy minerals, but this could hardly be appreciated in a geological environment which was very recently glaciated. Second, where local labor will sink a test pit for the equivalent of a dollar a foot or hand-steel a trench for a fraction of the cost of such work done mechanically in Canada, time-honored custom in industrial lands had better be very carefully analyzed before being heedlessly introduced into developing countries.

Ill. 2.2. United Nations team members George Kent (Canada) and Shiferaw Demissie (Ethiopia) panning for gold in the Manghi River, Ethiopia. This "primitive" method still works very well. (Credit: United Nations/Nagata/jr.)

Finally, when bilateral work in mineral exploration is on a loan basis, the temptation to adopt a massive, saturation approach must be avoided. Any particular program should be scaled to the total development needs of the country concerned in order to ensure that available assistance is spread across the most effective spectrum rather than applied in an unbalanced fashion.

The Piecemeal Approach

On occasion, foreign aid is provided for phases which are so preliminary that the end product cannot support economic development decisions. Again, I will refer to airborne geophysical surveys. Where these constitute the entire aid package, the recipient will often be confronted with a mass of data which is valueless until other help can be obtained for the ground follow-up required. Sometimes this may not be forthcoming within a reasonable length of time and then not in the needed quantities and quality. I therefore propose that mineral exploration work should constitute a coordinated whole aimed at reaching the point of indicating economic mineralization. This will usually mean that the various stages will have to be applied to a relatively modest geographical area. Instead of covering hundreds of thousands of square kilometers with closely spaced flight lines, prior study should be aimed at selecting a few tens of thousands which appear the most favorable in the light of existing knowledge.

The High Cost of Expertise

Where bilateral mineral exploration programs are financed on a loan basis, governments often get no bargain. The foreign expert can be very expensive. In fact, he can cost about three times as much as he would make at home. The base pay is a minor item. What greatly inflates the cost are the various allowances, subsidies, overheads and backstopping charges. One has to be shocked at such boondoggles as clothing allowances, particularly for geologists who would hardly go into the field sporting Hermes ties and Brooks Brothers suits. That such allowances do exist implies that in the home country clothes are not de rigueur.

On occasion, the heaviest burden of cost results from the high proportion of overhead employees on many aid missions. They usually far outnumber the producers, the people who get into the field and actually collect data. Often, particularly in the more advanced recipient countries, too many of the administrative staff are not as well qualified for their job as nationals who could be recruited for a fraction of the cost of the foreigner.

Tied Purchasing

Most donor governments dilute bilateral aid by barring or limiting international competition in regard to the provision of supplies, equipment and skills. Some programs have amounted to little more than providing outlets for home manufacturers.

The monopoly approach is also serious in regard to manpower. The donor nations being rich, their nationals who go abroad have to be very well paid indeed; and they are not necessarily very good. With due apology to a host of able and dedicated people who have toiled in the fields of foreign aid, one has to face the fact that, in general, the most capable man tends to stay close to the seat of power - at least on the edge of the limelight - rather than disappear for years into the obscurity of some far land. In this context the United Nations has the great advantage of international recruitment. When the boom in exploration during the past few years in Canada, the United States and Australia made it very difficult to interest younger Canadian and American geologists in United Nations work, the slack was taken up elsewhere - from countries such as the Latin American nations where there are more than a few well-trained and experienced geologists suffering from serious underemployment, and from nations such as Yugoslavia and Czechoslovakia where public agencies actively encourage their technical people to acquire foreign experience.

The "What's Good for You" Syndrome

The developing nations are too frequently sold programs which they do not really need in the context of sound priorities. Sometimes they have projects foisted upon them which they do not even want; but, as they seem to be free gifts (which they never are), they are accepted. In this matter, it must be confessed that the United Nations Development Program has also been occasionally guilty. By no means can all of the multitude of projects stand too close an examination as to practicality and desirability.

CONCLUSIONS

Despite the foregoing, I believe with total sincerity in the vital necessity of not only maintaining but greatly increasing foreign aid. One would simply like to see it done better; in fact, it may have to be, at least in the field of mineral exploration. There is no question that the bulk of new mineral supplies of the future will have to come from developing countries. Anything that increases efficiency in the search for these minerals should be very welcome.

3 Utilization of External Know-How in the Developing Countries*

THE NEED AND THE PROBLEM

Since World War II, one of the major preoccupations of leadership in the industrialized nations has been to do something to alleviate the tremendous imbalance between the haves and the have-nots. Hence, foreign aid which, 30 years ago, was largely the preserve of missionaries and those dedicated to helping disaster victims, has become a major plank in foreign policy. In part, motivation for helping has been altruistic - a slight stirring of conscience. Greater impulse has stemmed from the realization that people who are undernourished and clad in rags cannot create a market for the surplus production of the Western world but, instead, constitute a sickness which can become epidemic in a world where development of communications may not have made all men brothers but has greatly increased the incidence of rubbing elbows.

Now, after some 20 years of effort aimed at building up viable economies in underdeveloped areas, it can be seen that in relative terms the rich get richer and the poor get poorer. Even in any absolute sense, some countries are in a worse state than before World War II. Admittedly, the amount of foreign aid which has poured out has hardly been of sufficient magnitude to force the taxpayer to bypass the pub on his way home. However, it is not certain that a vast increase in the resources applied to development in the less fortunate parts of the world could make an impact even moderately proportional to the increase in magnitude. Without seasoned leadership enjoying some security of tenure, without fiscal stability (something very difficult to

---

* Adapted from an address to the Joint Meeting of the Engineering and Scientific Manpower Commissions, New York, May 20, 1965.

maintain in countries too highly dependent upon international primary commodity markets), and without a reasonably strong force of artisans and semiskilled workers, the superhighway which is donated today becomes a jungle track tomorrow.

A vital element in the total picture of foreign aid is manpower, particularly highly skilled manpower. Here the record is quite negative. It is more than probable that in terms of true functionality, the North Atlantic Community has extracted more technical competence and know-how in the postwar years from its poorer neighbors than it has returned in staffing aid programs. The industrialized nations have never known any McCarran Act (1) where capacity is concerned. A fundamental dynamic of their growth has been a relentless search for ability of whatever origin. This process is continual and, accordingly, there is a constant drain of professionals who can be ill spared from the underdeveloped parts of the world. North America and western Europe constitute a lure which is too often irresistible to the engineer, the scientist, the physician, even the highly skilled artisan. And one would make a mistake in assuming that this is simply a function of the paycheck. Probably more important is the freedom in research and operations which the above-average man enjoys in industrialized societies, and the chance to grow to the limit on one's capacity, which is frequently denied in less developed regions where, too often, the only source of employment is a government bureau which is grossly underbudgeted.

The bright young man or woman in Calcutta who develops an interesting empirical equation on air turbulence ends up with Boeing or Lockheed. A large percentage of foreign students never go home again; thousands of the most skilled people emigrate from Argentina alone every year. In return, the less fortunate parts of the world are flooded with advisers. A considerable percentage of these are dedicated and highly capable; others are adventurers who seek a new environment, preferably one which is not too competitive. However, no matter what the level of competence, the contribution of the adviser only rarely has considerable impact for the following reasons:

1. The average mission is short term - one to two years. Therefore, a disproportionate amount of time is spent in the vitally necessary but nonproductive business of getting one's feet on the ground. Even when the customs of the country and the language are known, many months have to be spent in establishing working relations, in being accepted and listened to.

2. Too many countries are "overexpertised." Some remote and somnolent country suddenly comes into the headlines and is invaded by a horde of advisers under bilateral or multilateral aid from governments, institutions and foundations. Rarely will the government officials concerned be able to sort this out to form a coherent, coordinated whole. Too often it will be a case similar to Stephen Leacock's knight who dashed from Lady Mary's boudoir, vaulted on his horse, and rode off in all directions at the same time.

3. The files of all ministries in all of the emerging countries tend to groan with reports, many of excellent quality, which will finally fall prey to termites because lack of infrastructure, capital, and political stability simply does not permit implementation. Of course, too often this represents a common failure of advisers - the inability to set sights low, aiming at the possible rather than the ideal.
4. There is probably too much advisory service and too little grass-roots effort.

SOME POSSIBLE SOLUTIONS

The more practical approach to providing technical skills to developing nations is exemplified by the Special Fund of the United Nations, now called the United Nations Development Program (UNDP); by that quite impressive group called the Peace Corps; by bilateral programs which involve such things as putting down water wells in villages, setting up a dairy, providing a visiting professor who can communicate; and so on. In this connection, the question has been raised: What can be done by associations of professionals in this area? Quite frankly, I am not optimistic; I do not think that they can do a great deal. A 20-year record of failure to narrow the gap between the underprivileged and ourselves is not going to be redressed even if the excess energy of every technical society in the industrialized world were consecrated to an all-out effort, a crusade. However, the battle to make a world in which the common lot is not misery is not going to be won by some great and glorious action. Instead, it will have to be a function of innumerable skirmishes - an inch gained here, a slight leak plugged there.

How then, and in what respects can the technical brains of the industrialized countries be more usefully applied to helping the underdeveloped countries to move ahead? In the public sector, I can only speak in regard to the United Nations.

I wish to point out first that a multilateral organization operating in the technical assistance field has some clear-cut advantages which should not be overlooked. It is free of the taint of self-interest with which, quite often unjustly, bilateral programs are burdened. In staffing it has access to brains and abilities from whatever part of the world. And, after years of financial crises, the United Nations has perforce developed a high capacity for getting a lot of mileage from its dollar. We fly tourist class!

Within the field of mineral exploration, with which I am particularly familiar, the United Nations has developed a roster of technicians and contractors capable of handling the problems involved in searching for what might be called the common minerals, usually, metallics. However, with respect to nonmetallics - several of which a man can specialize in during an entire lifetime - often the capacity for sound evaluation simply does not exist within the organization. In such cases, to appeal to a private company for consulting aid can prove fruitless because the technical assistance projects are often viewed as a mechanism for setting up competition. This is extremely narrow-sighted because, in the long run, mineral discoveries of importance made by the United Nations in nonsocialist countries will inevitably be brought to the extractive stage by those major concerns which have the capital and know-how to do so. In fact, the main purpose of the projects is to attract such capital by providing interesting and functional data.

In any case, it would be highly useful if, when the need for a specialist arises, the United Nations could approach a professional society, such as the Engineers Joint Council, (2) rather than private companies which may have difficulty in seeing the broad picture. If such organizations could develop a roster of specialists, drawing on both the public and private sector, from which people could be made available for short-term missions to solve specific problems, it would not only be of inestimable value to the United Nations but also it could serve as a source of industrial capability which governments could approach directly.

Several western European countries - Finland, Norway, Denmark, Sweden, Holland, Belgium, the Federal Republic of West Germany and France (3) have programs for hiring young nationals upon graduation from university and seconding them without charge to UNDP projects where the individual's particular discipline fits. Thus, these young people gain invaluable experience and provide usually quite functional additions to the team. I am sure that educators can readily appreciate what an excellent finishing touch could be provided to the graduate by, perhaps, two years in the Andes working with a group which includes economic geologists, geophysicists, geochemists, photogeologists and diamond drillers. It should be worthwhile to investigate the possibility of initiating a similar program in the United States, financed by the federal government or some foundation. Finally, the rather grandiloquent title of "associate expert" has been conferred upon these people. By and large, they have earned it.

Someday, it would be interesting to see one of the foundations or other providers of vast sums go into the translation business. One of the great barriers to achieving a high level of technical education in many parts of the world is the lack of teaching material published in the language of the country concerned. In Latin America, too high a percentage of the text books for the student of geology, engineering and medicine are in English. During their university life, the students manage to struggle along with these, but this can never be satisfactory. It is tough enough to master a scientific subject without having to cope with linguistics. By extension, periodic translation of scientific articles of wide-ranging interest could serve a very useful purpose.

Now, let us turn to free enterprise. For the largest part of the world as now constituted, private capital has to be by far the most dynamic mechanism of development. And whether capital in the industrialized countries realizes it or not, it is my contention that, for its own interest, it has to go abroad. This is not just because of the need of securing raw material sources, but because the economic stakes are commensurate with the tremendous challenge. If a five percent rise in the real income of the inhabitants of India is translated into pots, pans, shoes and transistor radios, the market implications are staggering. In responding to this challenge, industry is advised to bear in mind the following:

1. The day of the sahib and bwana has passed. This should be very well appreciated because the soundest business venture can be wrecked by an adverse impact upon the emotions of sensitive peoples. Such things as maintaining the upper management levels as the exclusive preserve of the nationals of the investing country not only overlooks the surprisingly strong reservoir of capacity which exists or can be developed in most places, but also creates resentments which can lead to newspaper headlines. With all the hard-won knowledge gained during these past two decades of how to live with our poorer neighbors, it seems incredible that in some instances separate payrolls are still being maintained, whereby the national employed by a foreign concern has a salary which is not only pegged at a low level but which bears no relationship to responsibility - the old business of the mine superintendent being on the "peso payroll" and making only a fraction of what his gringo subordinates get.

2. Sound policy when establishing an enterprise in a new country would be to endeavor to have the foreigners work themselves out of a job by neglecting no steps to develop the required managerial capacity with national employees. This is not only advisable psychologically, but also has certain very attractive economic features. The services and bonuses which are required to keep an employee in a foreign country can be staggering at times. The man costs ten thousand dollars a year in Pittsburgh, yet double that amount will no more than suffice to set up him and his family in New Delhi or Manila.

3. Private enterprise in the industrialized countries should weigh carefully the implications of hiring some bright youngster from, for instance, Turkey or India, who has just won a good degree from Harvard, Oxford, or the Ecole Polytechnique. It is a very human instinct to give a helping hand in such cases, but too often this will be mistaken kindness. That young person is needed far more in his own country and, in fact, in many cases he will have contracted to return in exchange for having had part or all of his education abroad paid for by his government.

4. The deep-seated prejudice toward becoming a business associate of a government bears re-examination. Facing the facts of life in this decade, there can be certain attractive features to a marriage between private enterprise and a federal government. The former delivers capital and know-how, while the latter puts up the opportunity and smooths the road to achieving an economic entity. That excessive nationalism which creates such a great barrier to foreign investment in so many parts of the world can hardly swing

into full stride when one of the partners is the country itself. Admittedly, such arrangements are difficult to make, but there are examples of where they have worked and worked very well indeed.

5. Finally, it is important that the industrialized nations take a hard look at the people they send abroad. Ideally, one could seek the extraordinary man but these are too few and far between. The least that should be aimed for is "the honest workman" - no genius, nobody who is going to revolutionize economies, but somebody who will see his job as far more than "nine to five and off we go to the first tee." Quite frankly, I cannot see private enterprise seconding very good men except where there is a clear vested interest. And, in government departments, there will always be a tendency to nominate "good old Joe," who really is not very useful but is a nice chap. In the long run, if too many second-raters end up in the underdeveloped countries, it can have ruinous effects, the waves from which will flow high upon the shores of the industrialized countries. There is very real self-interest involved here, and yet, I am not optimistic about the chances of improvement unless professional associations intervene, put pressure on governments to establish minimum standards, and take an active role in the recruitment process. There is no lack of precedent in this matter. In almost all provinces in Canada (and in most if not all states of the United States) a person has to be a member of the Professional Engineers to write an engineering report. Perhaps thought should be given to setting up international standards.

4 A Case for Greater Involvement of the International Organizations in Third World Mineral Development *

THE CHANGING SCENE

The mining industry is in trouble, real trouble. It has never been an easy, secure business. But today it is engulfed in revolutionary currents. No longer can we rely upon the checks and balances of market forces. Supply and demand largely dictated the economics of business enterprise in simpler days. Today politics, in particular nationalism, are increasingly dominant. And, as governments expand and put out tentacles of interference in the direction of the mining industry, we see far too little understanding of how that industry works.

Then, there is something which is relatively new - "stagflation": as the prices of minerals fall, the capital investment required for new extractive facilities soars at rates far greater than those cited by governments to indicate the magnitude of inflation. To illustrate:

- In 1971 it was calulated that $355 million would be required to bring the Cuajone property in Peru into production; five years later that stage was reached at a cost of $680 million.
- At the Caridad mine in Mexico, estimated capital investment doubled between 1971 and 1976.
- The requirements calculated for Cerro Colorado in Panama have risen in two years from $560 to $1,000 million.
- The budget of $240 million for Sar Chesmeh in Iran in 1971 had become $1,400 million by 1977.

---

* Based on a paper presented at the 1977 Annual Meeting of the American Institute of Mining, Metallurgical and Petroleum Engineers (AIME), Atlanta, Georgia, March 6-10, 1977.

- In about 1970, the best estimates of the cost of the Alaska Pipeline amounted to a sum in the neighborhood of $1,500 million. When it is completed some time this year (1977), the price tag will be about $8,000 million. This enormous overrun means that oil which can be produced for $0.50 per barrel will have to bear a pipeline transportation charge of $4.00 to $5.00 per barrel. Also, as legislation has been enacted whereby the pipeline has to be removed after the oil field is exhausted, further costs of about two billion in today's dollars can be anticipated.
- Another significant indicator can be provided by the diamond drill. The United Nations is a heavy purchaser of such equipment, spending an average of over a million dollars annually. Between 1970 and 1976 the cost of a specific package - machine, auxiliaries and spares - trebled. Over the short span of 1974 to 1976 it doubled.

All of the mines referred to above are porphyry coppers. They have been or are being brought to the production stage with the know-how of some of the most experienced firms in the world. None are likely to bear the heavy capital cost which would be required for environmental reasons if they were located in the United States. Besides, it would be a mistake to not anticipate additional overruns. In fact, with two of the mines cited, these are likely to amount to several hundred million dollars.

As capital costs rose out of sight, minerals prices have, on the average, languished over the past few years, not to mention the historical decline in the real price of many minerals throughout this century. The question arises as to why new mines are continuously being brought in. This anomaly can be explained principally by the momentum which builds up by the time the initial financing is secured. It becomes almost impossible to turn back even when it becomes evident that far more money will be required. In other cases, to call a halt to development will lead to expropriation.

For the next ten or fifteen years, it can be expected that in large parts of the world the established system will endure; that is to say, the major international mining houses will be the prime movers in providing the extra capacity required, unless of course there is a sustained depression, in which case all crystal balls shatter. Yet, a lot of major adjustments will be necessary. One of these should involve patching up what is usually an unhappy marriage between producers and consumers. It is clear that the trust necessary for an improved partnership cannot be developed in a climate of extreme demand and price fluctuations. Unfortunately, one finds it difficult to be sanguine about the prospects in this area. There has traditionally been an adversary relationship which today shows more signs of waxing than waning as indicated by the opposition of the industrial world at the fourth United Nations Conference on Trade and Development (UNCTAD IV) in Nairobi to the establishment of buffer stocks or other mechanisms for the purpose of moderating price fluctuations.

Considerable improvement in the geographical distribution of production will become mandatory sooner or later. Eventually, there has to be a limit to the inferiority of the material being mined. It makes for poor global balance, costly to all, to extract copper grading less than 0.4 percent in the so-called "safe" countries when far richer, larger and better-located deposits lie neglected in the developing world, particularly as the safe country of today can bare very large fangs indeed by tomorrow.

It is an understatement to say that the continued concentration of mineral investment in the four rich countries depicted in Table 4.1 is not desirable. If we accept the easily defended premise that the truly important mines of the future will mostly be in the three-million-tons-per-year or larger class, the dominance of the countries cited in Table 4.1 will become even greater unless accommodations are made and chronic imbalances corrected, at least in part. However, without a concerted effort being made to find new approaches, it is difficult to visualize significant change. The newly independent countries, that is to say, most of the people of the developing world, will frequently continue to view approaches by the international mining houses as a species of neo-imperialism. They will be prone to overestimating the cards they hold and political capital will be made by those who pose as staunch defenders of the national patrimony. Meanwhile, the mining houses will be very cautious, and the bankers exceedingly demanding. In effect, the impasse will continue in too many places in this world.

What can the international organizations, the United Nations and the World Bank in particular, do? How? Why?

THE UNITED NATIONS DEVELOPMENT PROGRAM

The United Nations first commenced undertaking mineral exploration in 1960 when a project started up in Chile which was financed in part by what was then called the Special Fund - the predecessor of today's UNDP. Up to June 1976, 184 projects had been completed, were in operation or funds had been earmarked for them. The sums involved here amounted to about $263 million, of which 53 percent has been contributed by UNDP. I have described and quantified the record in great detail elsewhere (see Chapter 5) and I do not need to repeat that here. It suffices to note that about two dollar's worth of minerals have been found for each cent spent or committed by the program.

There is considerable justification to argue that the industrial nations and the international mining fraternity should support in a material sense the expansion of the UNDP's mine-seeking efforts. However it would be utopian to assume that national mining companies, with neither the capacity nor the desire to seek properties abroad, would welcome the expenditure of their tax dollars to develop competition.

Today, when the incidence of mineral surveying abroad by the

TABLE 4.1. Analysis of Mines Producing More than 150,00 Metric Tons of Ore per Year

	Total (tons/year)	15,000-1 million (tons/year)	1-3 million (tons/year)	Greater than 3 million (tons/year)
OPEN PIT				
Number of Mines				
United States	118	43	30	45
Canada	46	4	21	21
South Africa	22	9	7	6
Australia	28	8	9	11
Subtotal	214	64	67	83
Rest of the world	265	127	90	48
Total	479	191	157	131
Percentage				
United States	25	23	19	34
Canada	10	2	13	16
South Africa	4	5	5	5
Australia	6	4	6	8
Subtotal	45	34	43	63
Rest of the world	55	66	57	37
Total	100	100	100	100
UNDERGROUND				
Number of Mines				
United States	66	43	17	6
Canada	107	73	31	3
South Africa	22	9	7	6
Australia	24	19	4	1
Subtotal	219	144	59	16
Rest of the world	319	274	89	28
Total	610	418	148	44
Percentage				
United States	11	10	11	14
Canada	17	17	21	6
South Africa	4	2	5	14
Australia	4	5	3	2
Subtotal	36	34	40	36
Rest of the world	64	66	60	64
Total	100	100	100	100

companies is generally at a low ebb, far greater funding and other support of UNDP mineral activities make a lot of sense. Despite the gloom and despair voiced by The Limits to Growth (Meadows et al., 1972), there are mines out there waiting to be discovered and, in my opinion, there always will be.

The UNDP has quite a lot to offer:

1. Transferral of risk. If nothing is discovered, the slate is wiped clean. With the majority of projects, where something is found, the United Nations will not receive repayment. However, of very recent years, additional financing by voluntary contributions has been obtained for the Revolving Fund for Natural Resources Exploration. When the Revolving Fund strikes it rich, it will be entitled to two percent of the gross value of production for a period of 15 years. Even if graced by great good luck the exercise will not be profit making. All revenues will be reinvested in mineral exploration in the Third World.

2. The track record has been good and, if momentum can be restored, it should be better. A constant process of selection going on for about 16 years had, until 1976, led to steady upgrading of the quality of personnel in the field, which had over 300 earth scientists at the peak. Sadly, in 1976 it became evident that UNDP was in serious financial difficulties. Worldwide inflation had caught up. The program was cut back everywhere. A considerable number of very good men were lost, some of whom had been on as many as four projects - that is to say with the organization for ten years or more.

3. A great deal has been learned over the years as to what will and what will not work. For example, it is now realized that in the early days there was too much emphasis upon airborne geophysics.

4. The United Nations has the enormous advantage of elbow room. Project areas of as much as a quarter of a million square kilometers have been put under a minerals reserve before the entry of the internationally recruited team. In a given geological province, the number of mineral deposits which can be found with today's concepts and tools will be roughly proportional to the area involved. This, in many places, rules out real grass-roots exploration by the mining companies, because they might have to negotiate for a year to get a concession covering a hundred square kilometers. And, too frequently, such concessions do not carry the automatic right to exploit any finds which may be made. That some mining companies will accept such a condition which, in effect, puts them totally at the mercy of governments, will never cease to amaze, unless it is accepted that some firms are more interested in speculation, stock market-wise, than in the "blood, sweat and tears" of bringing a mine into production. Perhaps too much has been made of modern exploration techniques. The finding and outlining of an important ore body is still an exceedingly expensive and accordingly slow process. But the process will be far more efficient if the search can be quickly focused upon the most favorable areas. For this task, the

combination of photogeology and geochemistry can be extremely effective. An area in the Cordillera of Argentina of about 143,000 square kilometers was reduced to about 5,000 square kilometers in one field session using these techniques without, as far as we know, overlooking anything of importance which could be found with the tools of the day. In short, the elbow room referred to above goes far to explain the record.

5. As to other support mentioned earlier, this would, in essence, amount to greater cooperation between bilateral donors and UNDP. The latter has always sought it, quite avidly at times. In part, this has stemmed from recognition that joint effort would increase overall efficiency considerably. More important, it is painfully obvious that bilateral assistance in the mineral exploration field has not met the near-term needs of the developing world. Insofar as the United Nations is aware, bilateral programs with funding vastly in excess of what the United Nations has enjoyed have not revealed the existence of a single major mine. Instead, effort has been concentrated upon development of knowledge in the traditional pattern of the donor country. Usually this has meant quadrangle mapping and training and investigations of scientific rather than economic orientation. There is no question that such programs will ultimately enrich most recipients. But if "ultimately" means 25 years from now, one can hardly feel satisfaction.

On an informal, more or less personal contact basis, the United Nations has received a considerable amount of ad hoc assistance from geological surveys of the industrial world: an age-dating here, seconding of a specialist for a short-term mission there, access to data, arrangements for fellowships, etc. At times, a team has been subcontracted from a survey for relatively extended periods. Nevertheless, on the whole, cooperation has been miniscule relative to potential. To illustrate, the Canadian International Development Agency has financed a great deal of airborne geophysics abroad - to the profit, it may be said, of Canadian firms. It is not quite so certain that the recipients gained much, because without resources to follow up such work, the end product amounts to no more than a map which has considerably less practical utility than an Exxon road guide.

To conclude, the United Nations is a very firm believer in offering what might be called a package deal from reconnaissance, through detailed work, to drilling records, leaving as few threads dangling as possible. When governments can only offer assistance covering a part of this process, if only for lack of qualified people willing and able to go abroad for extended periods, it would be highly advisable to investigate the possibilities of a joint program with the United Nations which would, in successful cases, give the recipient government the data and assistance required to negotiate follow-up by development capital. Where the bilateral source concerned is disinclined to accept a partnership, government authorities in the developing nations might be well advised to insist that a dialogue be opened and then monitor it very attentively.

THE WORLD BANK GROUP

The World Bank, either in its own name or through its subsidiary bodies, the International Development Association (IDA) and the International Finance Corporation (IFC), has provided more than $750 million in loans since 1957 for assisting mineral-related developments. Some of the loans have been on soft terms, but the great majority have involved rates and conditions consistent with maintaining the Bank as a going concern. Over the years, the World Bank Group's loans have amounted to about one-quarter of the total project costs. As will be noted in Table 4.2, Bank projects have been aimed not only at creation of mining and beneficiation facilities but also at a provision of infrastructure. Although some of the loans were made directly to governments, the majority were extended to incorporated bodies, either privately or publicly owned.

TABLE 4.2. World Bank Operations in the Mineral Sector through 1976

Country	Date of Loan Agreement	Amount of Contribution Millions U.S. $	Purpose
Botswana	June 1971	32.0	Shashe nickel project - infrastructure (town site, road, rail, water) facilities only
Zaire	March 1975	100.00	Copper mining and processing facilities
Gabon	June 1959	35.0	Manganese ore project - mine, cableway, railroad and port
Congo (Braz)	January 1967	30.0	Potash project - mining beneficiation, transport and port facilities
Guinea	Sept. 1968	64.5	Boke bauxite project - infrastructure (railway) facilities only
Guinea	June 1971	9.0	Boke bauxite expansion project
Mauritania	March 1960	66.0	Iron ore project - mines, railway and port
Mauritania	May 1968	20.0	Copper project - mining, beneficiation, town site and port

Table 4.2 - (Continued)

Country	Date of Loan Agreement	Amount of Contribution Millions U.S. $	Purpose
India	August 1961	35.0	Private sector coal production - mining and washing facilities
India	December 1961	19.5	Chasnalla coal mining projects - mining, washing and cable facilities
Philippines	July 1971	15.0	Nickel project - mining and smelting facilities
Israel	July 1961	25.0	Potash production - evaporation and material handling facilities
Greece	May 1970	8.6	Aluminum project - smelting facilities
Tunisia	October 1975	23.0	Phosphate mining and washing facilities
Dominican Republic	December 1969	25.0	Ferro-nickel project - infrastructure (power) facilities only
Bolivia	February 1974	6.2	Mining credit to medium miners
Mexico	February 1970	1.5	Iron ore project - mining
Brazil	January 1968	22.0	Aluminum project - mining, refining and smelting facilities
Brazil	August 1971	50.0	Iron ore project - mining, beneficiation and port facilities
Brazil	August 1971	46.0	Iron ore project - infrastructure (rail) facilities only
Brazil	April 1973	30.0	Nickel mining and refining
Chile	July 1957	12.2	Lota coal mine - mine equipment

Table 4.2 - (Continued

Country	Date of Loan Agreement	Amount of Contribution Millions U.S. $	Purpose
Chile	July 1957	9.6	Schwater coal mine - mine equipment
Chile	February 1958	2.2	Mantos Blancos copper mine - mining and smelting facilities
Chile	February 1970	10.9	Sagasca copper mine - mining and leaching facilities
Chile	March 1976	33.3	Copper sector project - mining and metallurgy facilities, rehabilitation of power plant
Peru	December 1976	40.0	CETROMIN - mining and beneficiation facilities
Guatemala	1975	15.0	Lateritic nickel mining and metallurgical processing to sulphide matte

<u>Totals</u>

	Number of Operations	Commitment (U.S. $ in millions)	Average Size of Commitment (U.S. $ in millions)
World Bank/IDA	20	683.0	34.2
IFC	8	103.2	12.9
Total World Bank Group	28	768.2	27.4

A great deal more involvement in mineral development by the World Bank would have infinitely more importance than expansion of UNDP activities. There is no intent to downgrade UNDP but, we have to admit that, for the near future, ensuring the development of known deposits in orderly relation to needs greatly transcends the requirement of discovering more.

The World Bank is by far the most powerful international financing institution. Few if any governments in the developing world could afford to play fast and loose with that organization. Accordingly, it is not difficult to see considerable merit in having the World Bank act as part of a mine development consortium, at least to the degree of supplying a part of the debt capital. This should ensure the signal contribution of an objective and powerful balance wheel for reconciling the differences between two parties when the gain of one has to directly lessen the take of the other. In the process, the very formidable resources of the World Bank would become deeply involved in the negotiating, planning, financing and technical appraisal facets of the project. Hopefully, these contributions would enjoy the confidence of the other parties. In fact, one might go so far as to say that in the years ahead if such inputs cannot be provided or accepted, generally there will be no deal.

With the World Bank in the lender role, private concerns would have very strong assurances that the sanctity of contracts would be observed at least until capital had been recovered. Also, it is more than probable that an institution dedicated to financial probity would ensure that sound practice was enforced in regard to depreciation, maintenance of working capital levels, and overall fiscal responsibility.

Bank participation would go a long way toward quieting the fears of the governments, the mining companies, and other lending institutions involved. Nevertheless, provision of funds on a loan basis would fall short of a far more attractive arrangement: sustained Bank involvement through acquisition of equity. The amount should be small - perhaps, five to ten percent of the shares - but it would guarantee that the concession agreement would be honored for its lifetime - unless, of course, there happened to be radical changes impossible to foresee which would threaten the existence of the enterprise if the status quo were maintained. Nothing is totally secure in this world, but there is no question that capital would sleep blissfully if in bed with what, by dint of a shameless steal, might be called the Young Lady of H Street Northwest.

There is a fair body of evidence that the Lady would welcome such a mating. In 1972, the Bank started to organize a meeting which was held in Washington in January of the following year. The participants - about two score in number - included many of the heads of the major international mining companies, public servants concerned with world-wide mineral developments, and bank officials. The intended thrust was telegraphed well in advance by a questionnaire which was sent to the Resources and Transport Division of the Department of Economic and Social Affairs of the United Nations, the executing agency for the great majority of UNDP mineral survey projects. The questions were very interesting, and in abbreviated form are listed below:

1. What is the availability of finance from non-World Bank sources? What is the order of magnitude of global investment in exploration?
2. What order of magnitude of additional funds might usefully be invested in exploration?
3. Is exploration activity in developing countries being held back because of insufficiency of finances?
4. What has been and what is likely to be the trend of foreign private investment for exploration in the developing countries?
5. Can adequate technical services for exploration work be obtained by developing country governments on reasonable terms?
6. Would World Bank lending to less developed country (LDC) governments for exploration help to strengthen the position of these governments at the stage of negotiating exploitation arrangements?
7. If the World Bank did lend for exploration, what would be the likely effect on the volume of foreign private investment in the mineral sector?
8. If the World Bank were prepared in principle to lend for exploration, should it concentrate on, or conversely rule out, assistance for any particular stage of the process?
9. What should be the criteria, apart from economic priority considerations, for selecting the countries or regions in which the World Bank would be prepared to finance exploration?
10. Where private capital can be had only on basically unfavorable terms, should the World Bank be willing to finance exploration?
11. Should the World Bank be willing to finance where private capital is excluded by public policy?
12. In the case of privately owned mining ventures, is there a need for World Bank or IFC finance at the exploitation stage?
13. In the case of mixed (i.e., government and private) mining enterprises, would LDC governments want to borrow from the World Bank for that portion of their share payable in cash? Would they welcome IFC investment?
14. What is likely to be the attitude of the private enterprise partners toward World Bank lending to LDC governments for this purpose?
15. Would World Bank lending to state-owned or state-controlled mining enterprises adversely affect the flow of private investment?

There were a number of other questions strongly indicating interest in diving into the breach, in doing something functional to resolve a global problem the seriousness of which has been obscured by the recession in the industrial world during the last few years. For the purposes of this chapter there is nothing to be gained by even listing these questions. The operative points were the evident interest of the World Bank in doing something about the "exploration gap" and in participating in the financing of mining developments.

It is clear that the World Bank knew a great deal about the often-cited exploration gap and was seriously considering moves which would, at least partially, close it. More important, the World Bank could foresee that there would be constantly increasing need for its participation in providing finances in the mining sector as governments became ever more nationalistic and private capital, accordingly, more cautious. Admittedly, no mention of taking equity positions was made but it takes no great feat of imagination to recognize that becoming a shareholder would be a very logical step, and in the end a mandatory one.

In any case, the Bank's attempt to increase participation in financing mining developments was, from all accounts, heartily endorsed by the participants in the January 1973 meeting. And then the whole affair seems to have died. Apparently special interests in the industry, the American mining companies with no foreign holdings, applied some heavy and effective pressure. This was a real pity. A union of governments, mining companies and the World Bank would have restored confidence in the viability of investment in developing countries. (1)

WHAT ARE THE CHOICES?

Let us consider a number of possible scenarios:

1. Continuation of the modus operandi which has evolved since the developing world started to flex muscles, namely, concentration of investment in the well-mineralized industrial nations. Such policy will not be workable much longer. The soaring cost of energy will sooner or later put an end to working deposits which are far lower grade than those in other parts of the world. In addition, there are other problems: constant ferment in the regulations of the supposedly stable countries; imposition of heavy royalties even when an operating loss has been incurred (British Columbia); impending expropriation of a segment vital to the world's mineral supplies by people without the slightest knowledge of the industry nor the capacity to manage the same (Saskatchewan); hostility to foreign investment while being heavily dependent upon outside involvement for development, technology and sales (Canada, Australia); ethnic struggles, the resolution of which may be far in the future but constantly cast a deep shadow upon current investment decisions (South Africa); and heavy, sometimes almost crippling but mandatory investment in pollution control (United States).

2. A holding of hands, a kissing of cheeks as the governments of the deprived world and the major international mining companies arrive at the point of giving full expression to their knowledge of interdependence. Everything which has happened in recent years suggests that instead of hands being held, fists will be clenched.

3. A brand new mechanism will become the norm rather than the exception: full government ownership, financing entirely by the international banking community and management and marketing by well-connected transnational concerns. This will be a decidedly regressive move. Despite traditional hyperbole, private enterprise is not normally a model of efficiency. That it does better than public enterprise is basically a function of the pressure from the owners as exemplified by the annual stockholders' meeting and the declared dividends. It is, therefore, not hard to visualize mediocre management when the operating concern does not have financial involvement and the personnel in charge can feel no dedication to a long-term career. In addition, indifferent performance will not necessarily lead quickly to beneficial change. The operating company will not take on the job without a three-to-five-year contract; otherwise, it will be difficult if not impossible to recruit people. Also, ineptitude is not prone to reveal itself and, therefore, a dangerous amount of time can pass before the rude arrival of the "moment of truth," the point of realizing that the operation is in serious trouble. Then, the really concerned parties, the government and the creditors, will be torn between paying heavy punitive damages to replace a concern giving indifferent performance by another which will initially be like a lost boy wandering the the wilderness and in the end may turn out to constitute no improvement. The interests of the owner - the government - will almost always be opposed to those of the enterprise. Particularly in the Third World, the demands upon the state are excessive in relation to resources and are always immediate. Accordingly, with any given enterprise, after debt has been retired or even earlier, the pressures on selected or appointed officials, whose expected life span in office is measured in months rather than years, can become irresistible. Instead of long-range planning, sound fiscal and operating policy, every force impels in the direction of maximizing the take from the mine, doing the taking now and forgetting tomorrow. In some cases this will mean high-grading, something which society cannot afford. In other cases it will lead to running down the plant and equipment with little provision for replacement, failure to keep exploration and development ahead of extraction, and a host of acts for satisfying immediate but unrelated needs, and let the Devil take care of the future.

4. A total logjam which impedes economic recovery, not to say growth: intransigent governments of the Third World; banking houses without the slightest degree of confidence in the ability of said governments to manage or even regulate any industrial affairs; mining companies

more prone to invest in timber lots or shopping plazas than foreign mining ventures.

5. Generators on order, transmission lines required, high demand for copper and aluminum, but not enough around. Industrial needs for silver exceeding output; Florida phosphate producers not capable of satisfying long-time clients. Who will come up with that which is required?

6. Finally, orderly development of mineral resources on a global scale because contracts are negotiated in good faith and their terms respected. When one considers the bewildering number of political philosophies and perceptions of self-interest existing today, one can argue that a "policeman" in the form of an international agency is vitally necessary to arrive at and ensure such a state.

5 United Nations Mineral Exploration Activities: 1960-1976 *

INTRODUCTION

Until the late 1950s, the efforts of the United Nations in the technical assistance field were modest and restricted almost entirely to providing advisory services. For a variety of reasons the program had limited impact. Recognition of this fact, combined with the marked increase in needs which became evident following the breakup of the colonial system, led to a major reappraisal of the role to be played by the United Nations. In the field of natural resources, it could be seen that standard service organizations - geological surveys, hydrographic agencies, administrative cadres, etc. - were being dismembered or disrupted and would have to be reconstituted or strengthened by one means or another. Above all, people would have to be trained to fill the gap created by the departure of the expatriate officials. From this ferment was born the Special Fund, now called the United Nations Development Program (UNDP).

In 1959, the first year of operations, 34 million dollars were earmarked by the Special Fund for 42 projects in a variety of fields. Recipient governments contributed an even larger total. By 1977, the annual budget of the UNDP had risen to about 500 million dollars, while counterpart contributions by governments almost certainly exceeded this sum.

The main thrust since 1959 has involved acting rather than telling. Instead of having an adviser to counsel a minister, teams of people - in this instance, internationally recruited earth scientists with their national counterparts - have been working on the ground, hacking through the jungle, climbing onto the hills or plodding through the

---

* Published with the same title in substantially the same form in Natural Resources Forum 1, no. 4 (July 1977). The copyright is reserved by the United Nations.

desert to develop the data required to identify ore bodies. As will be shown later, a considerable number of deposits have been found by these teams. In fact, for every cent allocated by the UNDP to its minerals program (which accounts for a rather modest portion, less than six percent, of the total UNDP program), nearly two dollars' worth of recoverable minerals has been outlined.

Virtually all projects concerned with mineral exploration and geological institution building or strengthening are executed, on behalf of the UNDP, by the United Nations Department of Economic and Social Affairs through its Office of Technical Cooperation, with substantive support and technical supervision provided by the Center for Natural Resources, Energy and Transport. Other specialized agencies of the United Nations are also active in certain areas of mineral sector development, though on a relatively smaller scale. In particular, the United Nations Educational, Scientific and Cultural Organization (UNESCO) and the International Labor Organization (ILO) assume responsibility for most of the training and educational projects; the International Atomic Energy Agency (IAEA), for uranium exploration; the United Nations Industrial Development Organization (UNIDO), for some technological research centers; and the World Bank, for preinvestment studies. A few of the projects are implemented by the UNDP itself, as a rule through subcontracts.

Ill. 5.1. A room with a view. Outside are members of the team with the United Nations assisted project in Cyprus. With his back to the camera is Gosta Tornqvist (Sweden) project manager. These are typical field headquarters.
(Credit: United Nations/PL/PAS.)

Ill. 5.2. Members of a United Nations team in an office tent deep in the jungle of Guyana. On the left is A.K. Banerjee (India) geologist. On the right, S.N. Saha (India) geophysicist. The third man is Gerald Moorhead (Ireland) project manager.

Ill. 5.3. Camping out in Guyana. (Credit: United Nations/RN/jr.)

The purpose of this chapter is to summarize the experience to date and evaluate, insofar as possible, the impact of the UNDP program in the field of mineral resources. It should be noted that the recently established United Nations Revolving Fund for Natural Resources Exploration is also administered by the UNDP, but its initial financing comes from special contributions by certain industrial countries. The Revolving Fund carries out mineral exploration and evaluation projects with technical support provided by the Center for Natural Resources, Energy and Transport. Since the Revolving Fund's operations only began late in 1976, they are not included in this historical review, but are discussed in detail in Chapter 6.

THE TIME FACTOR

Before tracing the United Nations record, a few explanatory notes regarding the time factor in mineral exploration are in order.

1. It is not always appreciated that the gestation period for mineral development is long and getting longer. Forty or fifty years ago, when the exposed, small, but relatively high-grade deposit was the norm rather than the exception, the time span between the beginning of prospecting and the commencement of production was often quite short - three or four years. Today, the vast majority of important new mineral finds have little or no expression at surface; grades are lower, and therefore, both the reserves and the required investment have to be correspondingly large. Inevitably, the principal phases involved in bringing a mine into production - prospecting, exploration, feasibility studies, negotiations with governments, financing, and development - take more time.

2. Perhaps too much has been made of modern exploration techniques. Basically, they are focusing devices rather than accelerators. As mineral exploration has moved more and more toward the detection of buried deposits, geophysical and geochemical methods have had two main functions: the elimination of areas where negative results largely rule out the existence of economic mineral deposits recoverable under current technology, and, conversely, the discovery of anomalous zones which may contain ore bodies. In the latter case, no technique developed in recent years is capable of significantly speeding up the exploration work required for quantitative and qualitative evaluations. Relatively laborious and costly investigations in the vertical dimension by diamond drill and/or underground workings are still required.

3. The typical UNDP mineral exploration project is located in remote areas with difficult access. The more easily accessible zones of favorable geology have usually been rather intensively

prospected. This factor not only slows down field work, but later, if a discovery is made, it seriously increases the financial burden and the time required for the construction of the necessary infrastructure.

4. Table 5.1 lists the major United Nations discoveries. (The definition of what constitutes a discovery will be considered later.) All of the deposits listed are believed to be exploitable in the foreseeable future. On the whole, the exploration record has been positive. But it would be remarkable if the only constraints to development were technical and economic. Instead, today, the principal determinant as to whether or not a potential ore body will come into production is the investment climate, irrespective of whether the country is developed or developing. In many instances such climates do not favor the growth of the mining sector. This assertion is amply supported by Table 5.1 which shows that only eight out of twenty discoveries are under active development or slated for development at present (mid-1977). The four that are in production have no great importance, collectively, in a relative sense. The effectiveness of the program as a whole has been seriously curtailed by the failure of many developing countries to attract investment capital in the mining field, and the likelihood of further deterioration is only too real.

Ill. 5.4. Expensive fuel delivery in Guyana.
(Credit: United Nations/RN/jr.)

TABLE 5.1. Time Lag in the Development of Major United Nations Discoveries

Country	Deposit	Commencement of Exploration	Commencement of Production*
Somalia	Wabo - uranium	1968	Unknown
	Mirig - uranium	1968	Unknown
	Dusa Mareb - uranium	1968	Unknown
Sudan	Hofrat en Nahas - copper	1968	Unknown
Mexico	La Caridad - copper	1966	1979
Malaysia	Mamut - copper	1965	1975
Chile	Los Pelambres - copper	1969	Unknown
Panama	Petaquilla - copper	1966	Unknown
	Botija - copper	1966	Unknown
Ecuador	Chaucha - copper	1969	Unknown
Guinea	Mt. Nimba - iron	1970	Unknown
Upper Volta	Tambao - manganese	1966	1981
Indonesia	Bangka - tin	1969	1975
Solomon Islands	Rennell Island - bauxite	1967	Unknown
Burundi	Gitega Area - nickel	1971	Unknown
Morocco	Barrechid Basin - salt	1971	1978
	Khemisset Basin - potash	1967	Unknown
Burma	Heinze Basin - tin	1971	1980
Togo	Marble quarries	1962	1969
	Limestone, cement plant	1962	1972

* Estimates for developments which are planned to reach the production stage after 1977 are based on data available in 1977

Ill. 5.5. Deluxe cruising in the Choco Valley of Colombia. Amidship is Derek Norris (United Kingdom), the United Nations project manager. (Credit: United Nations/John Littlewood.)

Ill. 5.6. Mule train bucking snowdrifts in the Cordillera in Argentina.

PROGRAM COSTS

The cost breakdown of projects in the minerals field which have been financed in whole or in part by the UNDP is shown in Table 5.2. Projects costing less than $100,000 have not been included. In most instances these involve only the provision of personnel in an advisory role. Also, the figures do not necessarily portray true, actual costs. In some instances they represent estimates, particularly with respect to contributions in kind, by the governments themselves or refer to expenditures authorized by the governing council of the UNDP but not yet disbursed. This is not too important in the present context, however, as my objective here is to show the geographic diversity of UNDP involvement and to indicate the order of magnitude of the total investment in the minerals field. Table 5.2 clearly demonstrates the wide-ranging nature of UNDP assistance, both spatially and technically. If there is any appreciable error in monetary terms, it is certainly on the side of understatement if only because in a number of instances complete and meaningful data on the government contributions are not available. Finally, Table 5.2 excludes projects concerned with the conversion of minerals to relatively advanced marketable forms. These are normally treated as industrial rather than mineral survey projects.

The UNDP and government expenditures and commitments are summarized in Table 5.3. It should be emphasized that even though my principal concern in this chapter is with economic factors, the establishment of training institutes and the strengthening of public organizations concerned with mineral development are also covered in the review. The contribution made by such projects cannot normally be evaluated in monetary terms, however. Yet, as I will describe later, in at least two instances where operations of such institutes were given a decided orientation toward economic geology, major mineral deposits were discovered in the course of their implementation.

As of June 1976, total UNDP expenditures and commitments amounted to $132 million of which about $100 million were for exploration and the bulk of the remainder for institution building and strengthening. Whereas government counterpart contributions were equivalent to about two-thirds of UNDP funds designated for mineral exploration, they were nearly double the UNDP contribution in the case of institution building and training.

MAJOR DISCOVERIES

In the sections which follow a number of discoveries will be cited. It must be emphasized at this point that the word "discovery" is used here with varying shades of meaning. At one extreme, discoveries refer to finds such as those of copper in Malaysia, Panama and Ecuador, bauxite in the Solomon Islands, oron ore in Chile, nickel in Burundi, and coal in

TABLE 5.2. Investment in Mineral Projects Assisted by the United Nations Development Program (a)
A. Executed by the United Nations (b)

	Mineral Exploration			Institution Building			Other		
	# of Projects	UNDP	Govern-ment	# of Projects	UNDP	Govern-ment	# of Projects	UNDP	Govern-ment
		(thousands U.S. $)			(thousands U.S. $)			(thousands U.S. $)	
Argentina	5	3669	6749	-	-	-	-	-	-
Banin	2	1001	702	1	498	255	-	-	-
Bolivia	4	4389(c)	2179	2	2043	2206	-	-	-
Brazil	1	103	-	-	-	-	-	-	-
Burma	4	5618	1441	1	213	-	2	1818	451(d)
Burundi	2	3218	304	-	-	-	-	-	-
Central African Republic	1	246	45	-	-	-	-	-	-
Chad	1	876	-	-	-	-	-	-	-
Chile	5	5377	3064	-	-	-	-	-	-
Colombia	5	3054	871	-	-	-	-	-	-
Congo	1	776	548	-	-	-	-	-	-
Costa Rica	1	124	81	-	-	-	-	-	-
Cyprus	1	1340	1595(e)	-	-	-	-	-	-
Democratic Yemen	1	125	-	-	-	-	-	-	-
Dominican Republic	1	2290	(c)	-	-	-	-	-	-
Ecuador	4	4015	836	1	987	575	-	-	-
Egypt	2	2969	3459	-	-	-	-	-	-
El Salvador	1	640	354	-	-	-	-	-	-
Ethiopia	1	1246	214	1	1440	4814	-	-	-
Gabon	2	667	242	-	-	-	-	-	-
Ghana	4	1493	390	-	-	-	-	-	-
Guatemala	1	877	697	-	-	-	-	-	-
Guinea	1	806	410	1	679	386	-	-	-
Guinea Bissau	1	149	134	3	445	144	-	-	-
Guyana	2	1530	1691	-	-	-	-	-	-
Haiti	2	1436	431	-	-	-	-	-	-
Honduras	1	1055	378	-	-	-	-	-	-
India	5	2905	1046	2	1795	1979	-	-	-
Indonesia	1	1080	2381	-	-	-	-	-	-
Iran	-	-	-	3	2292	6196	-	-	-
Israel	1	493	-	-	-	-	-	-	-
Ivory Coast	2	1743	850	-	-	-	-	-	-
Jordan	1	898	637	1	425	395	-	-	-
Kenya	1	671	409	-	-	-	-	-	-
Lesotho	4	2091	574	-	-	-	-	-	-
Liberia	1	821	74	-	-	-	-	-	-
Madagascar	1	1492	819(e)	-	-	-	-	-	-
Malawi	2	362	222	-	-	-	-	-	-
Malaysia	1	770	790(f)	-	-	-	-	-	-
Mali	1	173	-	-	-	-	-	-	-
Malta	1	317	69	-	-	-	-	-	-
Mauritania	-	-	-	1	387	347	-	-	-
Mexico	1	897	1850	-	-	-	-	-	-
Morocco	4	2726	3981	-	-	-	-	-	-
Nepal	-	-	-	2	1219	656	-	-	-
Nicaragua	1	738	355	-	-	-	-	-	-
Niger	2	1938	1034	-	-	-	-	-	-
Nigeria	1	566	491	-	-	-	-	-	-
Pakistan	2	1916	305(g)	-	-	-	-	-	-
Panama	3	1548	1071	-	-	-	-	-	-
Philippines	1	579	498	-	-	-	1	704	699(h)
Poland	1	995	2423	-	-	-	-	-	-
Republic of Korea	1	406	354	-	-	-	-	-	-
Rwanda	2	2175	403	-	-	-	-	-	-

Table 5.2 (Continued)

	Mineral Exploration			Institution Building			Other		
	# of Projects	UNDP	Govern-ment	# of Projects	UNDP	Govern-ment	# of Projects	UNDP	Govern-ment
		(thousands U.S. $)			(thousands U.S. $)			(thousands U.S. $)	
Senegal	3	1788	1083	-	-	-	-	-	-
Solomon Islands	1	898	634	-	-	-	-	-	-
Somalia	2	1444	1015(e)	-	-	-	-	-	-
Sudan	3	2179	958(c)	1	19	-	-	-	-
Swaziland	1	465	372	-	-	-	-	-	-
Thailand	1	280	134	-	-	-	-	-	-
Togo	1	1327	513(e)	1	370	170	-	-	-
Trinidad and Tobago	1	445	100	-	-	-	-	-	-
Tunisia	1	922	500	-	-	-	-	-	-
Turkey	2	1565	3682	1	467	111	-	-	-
United Republic of Cameroon	1	314	208	-	-	-	-	-	-
United Republic of Tanzania	2	961	547	-	-	-	-	-	-
Upper Volta	3	3135	1241(e)	-	-	-	-	-	-
Yugoslavia	1	438	277	-	-	-	-	-	-
Zaire	1	803	73	-	-	-	-	-	-
Zambia	1	831	746	-	-	-	-	-	-
Subtotals	123	95,184	59,534	22	13,459	18,234	3	2,522	1,150

B. Executed by the International Bank for Reconstruction and Development (IRBD - World Bank)

Country	Project Title	Project Cost UNDP	Government
		(thousands U.S. $ equivalent)	
Botswana	Preinvestment survey of utilities for mining development	398	557
Colombia	Survey of the Cauca Valley coal deposits	103	120
Surinam	Mineral survey	770	692
	Subtotals	1,271	1,369

C. Executed by the United Nations Educational, Scientific and Cultural Organization (UNESCO)

Costa Rica	Development of the Central American School of Geology	292	950
Egypt	Institute of Petroleum and MIning Engineering, Suez	1,661	1,288
Iraq	Petroleum Research Institute	1,043	2,569
Malawi	Geology	153	-
Saudi Arabia	Center for Applied Geology, College of Petroleum and Minerals, Jeddah	891	1,519
	Centre for Applied Geology, phase II	1,441	4,214

Table 5.2 (Continued)

B. Executed by the International Bank for Reconstruction and Development (Continued

Country	Project Title	Project Cost UNDP	Government
		(thousands U.S. $ equivalent)	
Spain	Higher Mining Engineering School, Oviedo	1,278	2,981
Sri Lanka	Geological Sciences	246	-
Zaire	National Mining Institute, Bukavu	1,743	4,096
	Subtotals	8,748	17,617
	D. Executed by the International Atomic Energy Agency (IAEA)		
Chile	Uranium prospecting	939	-
Greece	Exploration for uranium in central and eastern Macedonia and Thrace	364	226
	Exploration for uranium in central and eastern Macedonia and Thrace, phase II	357	676
Pakistan	Detailed exploration of uranium and other radioactive occurrences in the Siwalik sandstones in the Dera Ghaza Khan District, West Pakistan	659	-
	Exploration for uranium in the Siwalik sandstones	696	1,096
Turkey	Exploration for uranium in southwest Anatolia	747	601
	Subtotals	3,762	2,599
	E. Executed by the United Nations Development Program (UNDP)		
Afghanistan	Mineral evaluation assistance	131	10
Guatemala	Petroleum development	238	37
Kenya	Geothermal drilling	255	-
	Subtotals	624	47
	F. Executed by the United Nations Industrial Development Organization (UNIDO)		
Chile	Mining and Metallurgical Research Center	796	-
	Mining and Metallurgical Research Center, phase II	2,008	6,526
	Subtotals	2,804	6,526
	G. Executed by the International Labor Organization (ILO)		
Peru	Establishment of a center for vocational training in Recuay for miners	435	188
	H. Executed by the United Nations in Association with the Economic and Social Council for Asia and the Pacific (ESCAP)		
Regional	Southeast Asia Tin Research and Development Center	1,202	9,317

Table 5.2 (continued)

H. Executed by the United Nations in Association with the Economic and Social Council for Asia and the Pacific (Continued)

Country	Project Title	Project Cost UNDP	Government
		(thousands U.S. $ equivalent)	
Regional	Technical support for regional offshore prospecting in southeast Asia	696	115
Regional	Technical support for regional offshore prospecting in Southeast Asia, phase II	1,056	1,469
Regional	South Pacific offshore prospecting	159	148
	Subtotals	3,113	11,049

I. Summary

	Executing Agency	No. of Projects	UNDP Contribution (thousands U.S. $)	Government Contribution (thousands U.S. $)
A.	U.N.	148	11,165	78,918
B.	IBRD	3	1,271	1,369
C.	UNESCO	9	8,748	17,617
D.	IAEA	6	3,762	2,599
E.	UNDP	3	624	47
F.	UNIDO	2	2,804	6,526
G.	ILO	1	435	188
H.	UN/ESCAP	4	3,113	11,049
	TOTAL	176	131,922	118,313

(a) Approved between December 1959 and June 1976.

(b) The Department of Economic and Social Affairs, United Nations Headquarters, New York.

(c) Including one project financed by UNDP's Revolving Fund for Natural Resources Exploration.

(d) One project is a rehabilitation plan for the Dawdwin Mine and the other is for practical training in mineral exploration.

(e) Including a ground water component.

(f) Integrated river valley survey including hydrology, silviculture, soils and minerals.

(g) Including work in what is now Bangladesh.

(h) Practical training in mineral exploration.

TABLE 5.3. Expenditures and Commitments under the UNDP Mineral Survey Program: May 1959 - June 1976

Type of Project	Governing Council Commitments (a) (in U.S.$)	Government Counterpart Contributions (b) (in U.S.$)
Mineral Exploration	102,752,000	65,281,000
Training Institutes	8,334,000	14,865,000
Other Institutions	20,169,000	37,934,000
Rehabilitation of Mines	667,000	233,000
Total	131,922,000	118,313,000

(a) Including commitments for projects for which field operations have not yet started

(b) Understated, as complete and definitive data are not obtainable in a number of instances.

Bangladesh, all of which started from scratch, i.e., with no prior knowledge of the existence of these minerals. Then come deposits such as La Caridad in Mexico and Gaoua in Upper Volta - both copper - which had been the sites of small-scale, primitive, near-surface operations in the past. Of the same nature is a major porphyry copper deposit, Los Pelambres in Chile, which was first drilled by experts from the United Nations. That great mine finder, William Braden, had driven a couple of short adits on it half a century before and had given it up as the grade was decidedly uneconomic for that day. Some of the deposits listed, such as Tambao (manganese) in Upper Volta and Mt. Nimba (iron ore) in Guinea, were known long before the United Nations existed. However, the United Nations programs contributed materially by extensive drilling and other investigations, thus defining ore reserves, mineralogy and metallurgical characteristics. In the case of Tambao, UNDP-financed studies were also required to determine the location and economics of the railroad. At the other extreme are deposits which were known but whose development prospects were significantly improved through extensive work by the United Nations. The vast and unusually pure rock salt of the Berrechid Basin in Morocco was first encountered in holes drilled for petroleum; little, if any, importance was initially attached to this occurrence, however. Similarly, potash was known to exist at depth in the Khemisset Basin. The United Nations took on the rather formidable task of drilling these evaporites and financing feasibility and market studies. Extensions of known lead-zinc deposits in the Foussana Basin in Tunisia were also proved, though only on a relatively limited scale.

Mines in Operation

Malaysia

With expenditures of 82 million dollars, the Mamut porphyry copper deposit was brought into production in 1975 by the Overseas Mineral Development Corporation, a Japanese consortium. Based on 350 operating days per year, throughput has been designed as 15,000 metric tons of ore daily. The mine is an open pit, with a waste-to-ore ratio of 1.3:1. Planned annual production amounts to 27,000 to 30,000 tons of copper contained in a 25 percent concentrate, about two tons of gold and 14 tons of silver. Total reserves are estimated at 178 million tons with an average grade of 0.48 percent copper, sufficient for more than 30 years of production.

Togo

About six million dollars have been invested in opening up marble quarries. The bulk of the output is exported to Italy, a major producer of marble itself, which testifies to the superior quality of the product. The West Coast of Africa is a region deficient in limestone. Therefore, a major contribution was made by finding and outlining about 100 million tons of good-grade limestone. The deposit is suitably located, as it lies only 15 kilometers from Anecho, a port capable of berthing 100,000-ton vessels. The geological formation of the deposit is interesting and very advantageous. The limestone lies at a depth of ten meters or less and is immediately overlain by phosphate rock which can probably be exploited economically. Above that is a layer of bentonitic clay which will be required in the cement-making process. The initial investment in the mine and plant has been estimated at U.S.$43 million. Production in 1974 amounted to 130,000 metric tons of cement.

Indonesia

From 1970 to 1973, offshore from the Bangka alluvial tin deposits, the United Nations assisted in exploring for tin on the ocean floor at depths of up to 60 meters of water with a maximum penetration of 50 meters into the bottom sediments. More than 15,000 metric tons of fine tin were proved, all of which can be extracted by existing dredges belonging to the state-owned company, Perusahaan Negara Timah. Production has already started. A further 3,000 tons of tin can be extracted when dredges with deeper capacity become available. In addition, deposits containing almost 7,000 tons of tin were outlined and have been classified as probably and marginal reserves.

Assured Developments

As shown on Table 5.1, there are several major discoveries planned to come into production in the near future.

Mexico

The Caridad mine at Nacozari, east of Hermosillo, Sonora, is due to come on stream in 1979. A 60,000 metric ton/day concentrator will be designed to supply the smelter and refinery with enough mineral to achieve an output of 150,000 tons of copper per annum. Proved reserves amount to 734 million tons grading 0.7 percent copper and 0.016 percent molybdenum. Total reserves are probably several times this figure.

Morocco

Initially, the project was concerned with delineating potash beds first indicated by petroleum drilling in the Khemisset Basin. Fifty-five drill holes totaling 34,200 meters were completed and a number of technical and feasibility studies were carried out. A sylvinite zone containing 18 million tons grading 11 percent potash (K_2O) equivalent was outlined, as was a carnallite-enriched bed with reserves of about 240 million tons averaging 9.8 percent K_2O equivalent. As these deposits lie at depths of 500 to 800 meters, it is evident that grades are not high enough to support mining by conventional underground entry; the minerals should, however, be amenable to economic extraction by solution mining.

In later stages of the operation, 11,000 meters of drilling were carried out for investigation of rock salt in the nearby Berrechid Basin with considerable success. About 3,000 million tons were outlined in an area of approximately 15 square kilometers and with a thickness of about 80 meters. The salt lying at a depth of about 400 meters is exceptionally pure, averaging 98.6 percent sodium chloride (NaCl), with the bottom eight meters reaching 99 percent. Currently (mid-1977), a shaft is being sunk in preparation for production destined to satisfy three contracts for the development of Morocco's chemical industry. The aim is to produce annually 25,000 tons of chlorine, 28,000 tons of caustic soda, 25,000 tons of polyvinyl chloride and 500,000 tons of soda ash. Exports should be possible because the deposit is located within a few miles of the Atlantic port of Mohammedia.

Burma

In the Heinze Basin lying about 400 kilometers southeast of Rangoon drilling for alluvial tin offshore and in mangrove swamps produced data sufficiently definitive to result in a large World Bank loan to bring the property into production. This is the first instance of the World Bank following up a United Nations discovery.

Potential Developments

Ecuador

In the Chaucha Valley, which lies about 30 kilometers from the Pacific Coast, an intensive geochemical program and various follow-up surveys by United Nations in 1969 indicated a large porphyry copper deposit also containing molybdenum. Subsequently, a Japanese firm drilled the property with an initial commitment to spend almost five million dollars on detailed exploration. About 60 million tons were outlined, grading the equivalent of 0.7 percent copper when the molybdenum content was taken into account. The deposit was considered marginal, and the Japanese investor withdrew. It is likely, however, that the Chaucha deposit will excite interest again in the near future, particularly since the chances of substantially increasing the reserves appear to be excellent.

Panama

The discovery of copper-molybdenum deposits at Botija and Petaquilla in 1966 sparked the search for copper in that country on a wide scale. Subsequently, in the early Seventies, a Japanese company obtained an exploration permit covering both deposits which lie about four kilometers apart. On the basis of about 11,000 meters of drilling, measured and indicated reserves of the order of 300 million metric tons were outlined. The average grade is about 0.7 percent copper.

Upper Volta

The reserves of the Tambao manganese deposit are not large in terms of manganese content - about 12 million tons of oxides - but they are of a very high grade, averaging 56 percent. In addition, the deposit contains appreciable quantities of nsutite, a manganese mineral used in battery manufacture, which commands a price about three times greater than that of standard-grade material. The chances for production of chemical-grade manganese, also a premium earner, are good.

Equity in the mine will be split between Upper Volta, Japanese, German, United States and French capital in the following respective percentages: 51, 30, 9, 7 and 3. This group is currently engaged in a feasibility study to be completed in March 1977 to determine if a profitable operation is indicated despite the high cost of an access railroad which, including rolling stock, is now estimated at $120 million. It is interesting and starkly indicative of what has happened to the world economy during the past few years that a soundly estimated pricetag for this railroad in the late Sixties was only $35 million. If the feasibility study is favorable, the mining group will have to invest an additional $50 million to open up the mine. Annual output will probably be on the order of 500,000 metric tons and the start-up date 1981. The government is apparently prepared to provide its share of the investment. (1)

Solomon Islands

On Rennell Island, a United Nations mineral survey discovered a bauxite deposit and, subsequently, reserves totaling 30 million tons were proved. Nearby on Wagina Island a similar quantity was delineated by Conzinc Rio Tinto. Currently, feasibility studies are being undertaken by Mitsui Mining and Smelting with the objective of determining the viability of a plant producing about 1.2 million tons of alumina per annum. The investment requirements are estimated at about $300 million. So, presumably, the feasibility study will revolve to a considerable degree around the availability of capital on reasonable terms. If the results of the study are favorable, the plant may come on stream in 1980. (2)

Deposits Not Under Active Consideration

A number of potentially economic discoveries appear to lie neglected. These include uranium in the Mudugh Area of Somalia, copper at Hofrat en Nahas, Sudan (it is understood that a concession is being negotiated), the Gasua copper deposit in Upper Volta, Guinea's Mount Nimba iron, nickel laterites of the Gitaga area in Burundi, and Los Pelambres copper and Boqueron Chañar iron in Chile. The reasons for the lack of development vary, but political and financial factors seem to be at the root in most cases.

SUMMARY OF MEASURABLE BENEFITS

Tables 5.4 and 5.5 provide a reasonable reflection of the measurable results achieved to date by the UNDP mineral surveys.

As shown in Table 5.4 the capital investment which has been or should be generated by the program amounts to more than a thousand million dollars. Only developments which have reached the productive stage or are sufficiently advanced to be classified as assured or probable are included.

With regard to Table 5.5, which shows the quantity, grade and value of the ores discovered, a few explanatory words are necessary. The value assigned to a mineral deposit has been obtained by multiplying the units of metal in the ground by the current market price and adjusting downward (as will be explained later). Admittedly, this is a simplistic approach. However, in this instance a number of deposits with a wide variety and mixes of commercial minerals are being considered. They are located in different countries, each with its distinctive level of development - economic, political, and physical. In effect, the variables predominate and the constants are few and far between. Under such circumstances, it is manifestly impossible to obtain the parameters needed for more sophisticated analyses such as discounted cash flow.

TABLE 5.4. Reported Investment Incurred or Considered for United Nations Discoveries*

Country and Status of Project	Deposit	U.S. $ Equivalent
Completed		
Malaysia	Mamut copper	82,000,000
Togo	Marble	6,000,000
Togo	Cement plant	43,000,000
Indonesia	Offshore tin	15,000,000
Assured Developments		
Mexico	La Caridad	500,000,000
Morocco	Berrechid Basin - salt	182,000,000
Burma	Heinze Basin - tin	27,800,000
Probable Developments		
Solomon Islands	Rennell Island - bauxite	300,000,000
Upper Volta	Tamboa - manganese	170,000,000
Preinvestment Studies		
Ecuador	Chaucha - copper	4,700,000
Panama	Botija-Petaquilla - copper	4,000,000
Trinidad	Petroleum	1,300,000
	Total	1,335,800,000

* Even if some of the developments rated as probable a year ago have to be classified as doubtful now (in view of the financial crunch and depressed market outlook), the total still is valid because of the inevitable cost overruns with projects under development. (Ed.)

TABLE 5.5. The United Nations Mineral Exploration Record

Country Mine Principal Material	Reserves (metric tons in thousands)	Grade (percent)	Recoverable Mineral (a) (metric tons in thousands)	Value (b) (U.S. $ millions)
Somalia Mudugh Area Uranium	-	-	4,820	319 (c)
Sudan Hofrat en Hahas Copper	8,700 (e)	4.0	282	426 (d)
Mexico Caridad Copper	734,000	0.7	4,162	6,293 (d)
Malaysia Mamut Copper	178,000	0.48	692	1,046 (d)
Chile Pelambres Copper	248,000	0.78	1,567	2,369 (d)
Panama Petaquilla and Botija Copper	300,000	0.7	1,701	2,572 (d)
Ecuador Chaucha Copper	60,000	65	340	514 (d)
Guinea Mt. Nimba Iron	600,000	56	486,000	2,741 (f)
Upper Volta Tamboa Manganese	12,000	0.4 kg/m^3	9,720	793 (g)
Indonesia Bangka Tin	63 million m^3	0.22 kg/m^3	20	172 (h)
Burma Heinze Basin Tin	31 million m^3	0.22 kg/m^3	6	52 (h)
Solomon Islands Rennell Island Bauxite	30,000	45	24,300	364 (i)
Burundi Gitega Area Nickel	116,000	1.44	1,228	6,521 (j)
Morocco Berrechid Basin Salt	3,000,000	98.6	2,430,000	1,822 (k)
				26,004

(a) Assuming a ten percent loss in mining and ten percent in beneficiation. Where no beneficiation has been involved, ten percent has been allowed for shipping losses.

(b) At point of sale to end users except for bulk materials

(c) Based on $30 per pound uranium ($U_3O_8$).

(d) Based on United States producer delivered price of refined copper, February 1977.

(e) Reserves measured only to the indicated and inferred stage; they are considered admissible as the potential for considerably greater tonnages is high.

(f) Based on Lake Superior Old Range ore at lake ports as given in Engineering and Mining Journal, January 1977, discounted by 75 percent.

(i) Based on $15 per ton.

(j) Based on price of cathode nickel given in Engineering and Mining Journal, January 1977.

(k) Based on $3 per ton discounted by 75 percent.

Furthermore, it is not possible to apply "present value" concepts, because with many of the deposits concerned, there is no way of knowing when the shovels will dig and the rod mills will roll. In some cases it will undoubtedly be many more years hence than would apply if anything approaching a global mineral development strategy existed.

In addition, the gross value of a mineral in place cannot provide any precise idea of the benefits which will accrue to the country which owns the resource. The net revenue to the nation has to be reduced by the cost of supplies and services paid for in foreign exchange, international shipping charges, cost of smelting and refining (when done abroad), amortization of foreign investment interest payments and repatriation of profits. It must also be appreciated that the multiplier effects of mining activities in the developing world on the whole will rarely be of much significance. Modern mining of any importance is anything but labor-intensive, and in most cases the skills, equipment, supplies, and financial and marketing connections required must essentially come from the industrialized countries, in the short and medium term at least.

Nevertheless, in an average mining operation, once capital investment has been recovered, about 50 percent of the gross value of production will tend to remain in the country, even if it is a less developed country, for payment of labor, locally produced supplies, power, internal transportation, port loading charges, royalties (if and when these are imposed), and the government's share of profits, either from equity or taxes or both. As countries advance in regard to both the development of human skills and the production of goods and services, it can be expected that even with a mine which is wholly foreign owned, eventually between 80 and 90 percent of the proceeds will accrue to the nation, particularly if the scale of operations is sufficient to support local smelting and refining.

Some observations are required at this point to explain the approach used in assigning values to mineral deposits. The measure of value chosen as a starting point was the price at which the mineral enters the market for conversion into capital and consumer goods. Obviously, this price rarely approximates what will be paid at the point of export; yet, it is not inappropriate to select the value of the commodity concerned when it arrives at the point of transformation by manufacturing processes. Second, the product of the multiplication of contained mineral in reserves by the market price was reduced in all cases by 20 percent. This is reasonable, on the average, in dealing with minerals such as base metals where losses of that order can be expected in the mining and beneficiation stages. With bulk minerals such as bauxite and phosphate rock where no on-site beneficiation is involved, the same factor was applied because these low-value minerals are shipped in their natural form and are thus subject to significant losses in transit, whereas the more valuable minerals in concentrates are usually transported in containers such as bags and barrels. Finally, the values assigned to Mt. Nimba iron and Moroccan salt were discounted heavily, since serious distortion would otherwise have occurred. These two deposits are not likely to be exhausted for a hundred or hundreds of years. To assign full value does not make much sense when most of the other mining operations will tend to take place with fixed-term - say 25 to 30 years - exploitation rights. Under such constraint, every effort will be made to assign a life to known reserves more or less equal to the term of the contract - to mine as much as possible while the opportunity to do so exists.

It is obvious that the selection of criteria and methods for assigning a monetary value to mineral discoveries can alter the results significantly. The approach I have used is, I believe, conservative.

NONMEASURABLE BENEFITS

There have been other favorable results which are difficult to quantify but whose importance is not in doubt: training and practical experience, which work both ways by upgrading not only nationals but also members of the internationally recruited teams; the building up of a vast body of data and experience covering the widest spectrum in the history of mineral exploration which, if pulled together, analyzed and synthesized, can serve to guide mineral exploration throughout the developing world.

Valuable contributions have also been made by the type of United Nations project which is not aimed at discoveries but is designed, rather, to improve techniques and skills. As an example, the Mining and Metallurgical Research Institute in Bolivia has made significant advances in research on the volatilization of cassiterite and in increasing the efficiency of existing concentrators.

Another example is the work done by the Institute of Applied Geology in Manila. This institute was not designed to create new riches other than in the classical sense enunciated by Adam Smith: that the wealth of a nation is its people. However, in the process of training Philippine mining engineers in geological exploration, a great deal of practical field work was undertaken and, in at least two cases, this led directly to important discoveries:

1. At St. Nino, Mountain Province, the Baguio Gold Mining Company was exploring a large low-grade copper deposit in 1965. With the cooperation of the company, the institute conducted a considerable number of geochemical and geophysical surveys in the area. Previously, known reserves were about eight million tons of 0.4 percent copper. Finds made by the institute resulted in trebling this figure and in raising the average grade.

2. At Boneng, institute personnel discovered two mineralized cupriferous zones, one of which is being prepared for production. Unfortunately data are not available on reserves and grade.

In files all over the world there are project reports which refer to mineral occurrences for which no follow-up action is planned today insofar as we know: for example, Buck Reef gold in Tanzania, atapulgite in Togo, and carnallite in Morocco. Sometimes the reference in reports may do no more than point out a certain type of hydrothermal alteration, a lithologic-tectonic relationship, the mineralization revealed in a small outcrop, or certain regional connotations on airborne geophysical maps. However, these technical data may be of vital importance one day. As the geology of a region becomes better understood, attention can focus on a certain structure which was hitherto ignored; a strong demand may unexpectedly arise for minerals not actively sought today; and seemingly minor "blips" on geophysical profiles can take on new significance. This suggests that the process of re-examination should be almost continuous. Each new find that broadens geological knowledge tends to point out others. This may be limited in a spatial sense to providing no more than a clue to possible extensions of known deposits. At the other extreme, a single discovery may produce ripples which extend for thousands of miles.

It goes without saying that any well-conceived and well-executed scientific effort does not have to produce a saleable product to be worthwhile. Before the first nylon fabric was spun or the first sulfanilamide pill was swallowed almost certainly hundreds of abortive experiments provided the broad base which led to the successful results. Similarly, a certain proportion, always very high, of mineral exploration is bound to be fruitless in terms of outlining ore bodies. However, there is an important question to ask in order to judge whether or not a project has been useful: Has the work been done well enough in the light of existing techniques and knowledge to have provided a sound assessment of the mineral possibilities or their absence? A negative answer to this question permits reshuffling of priorities and reassessment of avenues for national development, and it dampens unjustified wishful thinking. Fortunately, in such cases only rarely will the decision to downgrade mineral exploration activities be of a permanent nature. Earlier in this account, I referred to those projects which have achieved positive results pointing to the proved, probable or possible existence of ore bodies. The word "success" is frequently used to describe such projects. This categorization is grossly misleading, however, because it implies that all others have been failures. This is simply not so if the work has been well performed and if documentation is preserved for subsequent reappraisal as circumstances change.

CONCLUSIONS

There is at present practically a hiatus in mineral exploration in the developing world by those mining houses which are capable of doing a good job and have the stature to carry a discovery through to production if conditions are right. Sadly, only rarely are conditions reasonably favorable and, accordingly, too often mineral exploration in the Third World is limited to visits by geologists from abroad who are empowered to look and report but not to invest in concrete work.

There is no question that an organization such as the United Nations has an important role to play in this area. When mining companies and governments are unable to reach agreements regarding the development of proved deposits which would normally be economic, it is manifestly unrealistic to expect the companies to undertake the high-risk business of exploration in what might be called terra incognita. It is suggested that the word terra should not be taken literally; it should be interpreted instead as covering a wide gamut from geology through to the inner chambers of government. The United Nations, on the other hand, is not particularly concerned with hazards that are quite separate from those stemming from the vagaries of geology. It does not have to publish annual profit and loss statements. Accordingly, in the unsettled world we live in, it is evident that an organization which enters at the invitation and with the active support of governments and adopts the approach of carrying on simultaneous investigations in scores of countries, is almost certainly best suited to undertake the very high risk work of "wildcat" exploration in environments which are essentially untested physically.

6 The Revolving Fund for Natural Resources Exploration: Truths and Misconceptions*

A great deal has been written in recent years about the exploration gap which has to do with the disproportionate amount which is spent on mineral exploration in a handful of industrialized nations as compared with outlays in the developing world. Geology not being defined by national frontiers, it is evident that this concentration is, in part, the result of the tendency of investment to favor the home territory - of the preference for the familiar over the relatively unknown. A more important and specific deterrent to investment is the volatility, the political and social turmoil, characteristic of many developing countries. The rising tide of nationalism which has become such a dominant feature of the twentieth century is well advanced even in some developed countries hitherto considered as havens for mining investment. Canada and Australia provide outstanding examples, with Ireland a close third. Yet, it is clear that unless the current global recession continues indefinitely - new mines must be found and brought into production constantly if only to meet the needs of an ever-expanding population. And, increasingly, unless we are reduced to mining almost barren rock in countries such as the United States, in the long term these mines will have to be located in developing regions.

GENESIS OF THE REVOLVING FUND

In response to this obvious need, the United Nations Revolving Fund for Natural Resources Exploration has come into operation. The Fund provides a source of financing mineral exploration additional to, and

* From an unpublished paper read at the Seminar on Mineral Economics held at McGill University, Montreal, February 23 - April 2, 1976.

separate from, the United Nations Development Program (UNDP) although it is administered, during the initial stages, by the UNDP. Its principal objective is to assume the very high risks involved in finding mines, from regional geological reconnaissance to delineating ore bodies - or better said, to prove up masses of mineralization which could be exploited economically under a just and stable political regime. The Fund is thus designed to provide a bridge over the most dangerous part of the mine development process.

One of the prime movers in the genesis of the Fund has been Japan. This country was the first to recognize that whereas a private mining company might have to negotiate for years to obtain a concession covering only a few tens of square kilometers, the United Nations suffered under no such restrictions and therefore appeared to be the best vehicle for undertaking wide-scale searches for minerals in the developing world. It was implicit that Japanese support of the Fund did not ensure it a favored treatment position in obtaining exploitation rights, the award of which is entirely the right and responsibility of sovereign governments. However, it did not escape the supporters of the Fund that, in a world with vast international trade in mineral raw materials, any new discovery of importance stood to benefit to one degree or another all consumers, regardless of who became the producer.

There is no need to go into details of the birth pains of the Fund which were very protracted indeed. At meeting after meeting over a period of years, the broad principles were hammered out culminating in the official creation of the Fund by a resolution of the General Assembly which was passed in December 1973. Another year and a half were required to develop the Operational Procedures and Administration Arrangements for the Fund. (See Appendix A.)

At the same meeting in Geneva in June 1975 which ratified the Procedures and Arrangements, the first two projects were approved.* Neither are operational yet (February 1976) because, for reasons which I shall explain in detail later, it has proven exceedingly difficult to negotiate project agreements. Anything which is new, different and constitutes a clear departure from established practice is not likely to find ready acceptance.

SPECIAL CHARACTERISTICS OF THE REVOLVING FUND

The Fund is distinguished from other types of United Nations development assistance in several respects.

1. It was designed to revolve, that is to say become self-sufficient

---

* By the end of 1978, 14 projects had been approved, of which four were later cancelled, five are still being negotiated, two have been completed and three are in operation or will be very soon.

financially. When one of its discoveries reaches the production stage, the Fund will be entitled to a repayment, hereafter referred to as the Replenishment Contribution, of two percent of the gross value of production for a period of 15 years from the commencement of commercial output. There is a limiting factor, however: all obligations of the government will terminate in 30 years from the start of any particular project. Accordingly, any discovery which is not exploited for, let us say, 20 years, will be subject to assessment of the Replenishment Contribution for ten years only.

2. There was also a decision to impose a maximum limit - a ceiling - on total Replenishment Contributions. Although sufficient data were not available to set such a limit, it was suggested that the ceiling could be in the region of 15 times the original investment of the Fund, in constant prices, and the door was left open for governments to negotiate a limit for specific projects. In effect, because of political expediency, diplomacy, cynicism and almost total ignorance of the economics and structure of the mining industry, the Governing Council, by introducing the ceiling concept effectively ensured that the Fund will never have a chance to revolve. This is probably not important if one thinks in terms of resources. If it finds ore bodies in the developing world, and it will, the Fund's existence will be more than fully justified. Who gains, who pays - these are minor items. What counts is enhancement of knowledge of our physical world, a world in which geology has transcendental importance. However the question arises: "Will nations continue to contribute when it is more than evident that recipients are resolved that the Fund shall not revolve?"

3. The Fund will pay all costs. Otherwise, there could not be full justification for obtaining the Replenishment Contribution. The Fund will, in most instances, spend a considerably larger proportion of its budget in national currency than has been the case with regular UNDP-assisted projects, which have suffered under the stricture that anything which can be paid for in local currency properly constitutes a contribution of the government. Thus, for example, contracting of local firms for specialized services has rarely taken place because the government simply cannot meet the cost. Instead, a foreign firm is engaged by the United Nations at a considerably higher cost; similarly, with personnel. The existence within the country of a highly skilled national has had no value whatsoever if the government could not meet his salary demands. The Fund will seek such people assiduously.

4. On the basis of past experience, it was calculated that the Fund might become self-sufficient - or, to use official terminology, may revolve - in about 20 years. While this implies that Replenishment Contributions will match expenditures, it is not envisioned that the Fund will ever become a profit-making operation because income will always be reinvested in further mineral exploration in the developing countries. Of course, the ceiling referred to above will make a mockery of all calculations if enacted.

5. The criteria which will determine the acceptability of a project will, in general, be more stringent than those applying to regular UNDP assistance. That organization has had to take on projects of dubious merit because of the obligation to achieve widespread geographical distribution and to take into account the plight of the least developed countries. Geology, however, is not the smooth spreading of butter across a piece of bread representing the world. It is very much an erratic phenomenon. Accordingly, the reference to equitable distribution and the least developed countries in the Operational Procedures and Administrative Arrangements probably amounts to no more than shopworn ritual and should not in any sense hamper the Fund from selecting projects which appear to be reasonable gambles.

6. The Fund initially will only search for hard minerals. Hydrocarbons and geothermal energy will be excluded in the early years chiefly because of the high cost of exploration for them. Ground water will also be excluded if only because in most cases its exploitation could not support a meaningful Replenishment Contribution. The United Nations would be somewhat chary of the concept of posting a tax collector at the village water hole.

7. Until more than ample funds exist, feasibility studies will not be undertaken. In the first place, these can be extremely costly. Secondly, as feasibility workd would usually revolve about deposits which have already been discovered and outlined, the regular repayment formula of two percent of the gross value of output over 15 years could hardly be applied. Therefore, a body of experience will have to be built up before sound replenishment principles can be established. Probably the only mechanism which will work will be to treat such expenditures as short-term loans - let us say, five years - bearing a relatively low rate of interest and to be repayable in funds adjusted for inflation. It is tempting to think in terms of the ultimate in soft loans, namely, one which is interest free; however, in recalling a history replete with examples of governments reneging on obligations, something on the order of eight to ten percent will no doubt be required if the Fund is to break even.

Objectively, the Fund, as described, represents a very good deal for several reasons:

1. The Fund takes all the risk. If there is no discovery, there can be no claim for repayment.

2. The Fund will undertake these risks in places where private capital does not dare to venture these days.

3. Whereas a private concern will rarely make a first commitment in developing countries for early-stage exploration amounting to more than, let us say, a quarter of a million dollars, the Governing Council of the Fund has already authorized a project for which the first phase cost over $1.2 million.

4. At the very beginning of the operation, a total of over $11 million has been paid in or pledged for 1976 by the following countries: Japan, Canada, the Netherlands, Belgium and Iraq.* Henry Kissinger's speech, read by Patrick Moynihan to the General Assembly in September 1975, accorded warm support for the Fund by the United States, and a sizable contribution is expected for 1977.

5. As noted before, the Fund aims to bridge that high-risk area which the multinational mining companies are not prepared to cover in many parts of the developing world. However, it would be folly to believe that an important mining enterprise can be brought into being today without major participation of such companies except in countries having centrally planned economies. In fact, even these are increasingly making overtures to Western nations for assistance in industrial development.

6. The Fund will require a two percent Replenishment Contribution for discoveries stemming usually from the highest-risk stages of exploration. Important private capital today will not take on a mining venture even though it revolves about a proved ore body unless the most conservative calculations indicate an annual payback of 20 to 25 percent of the gross value of production before interest, amortization and taxes. In actual fact, they cannot. The scale of investment in this inflation-bedeviled age is such that by far the largest part has to be of a debt nature. Bankers do not survive unless they subject discounted cash flow figures to very hard scrutiny. In effect, no mining company could stay in business if it undertook exploration without the chance of a return several times greater than the Fund will exact.

7. From past experience of the United Nations and of the mining industry in general, it can be predicted that from the commencement of the search for minerals in relatively unknown areas to the production stage in successful cases, 12 to 15 years will elapse with large mines. Fortunately, the Fund will constitute an interest-free organism. The donor nations are not setting it up as a business.

8. Finally, the UNDP can no longer be viewed as a relatively affluent source of mineral exploration funds. For a variety of reasons which do not warrant discussion here, the UNDP is insolvent. (1) (No state secrets are being divulged because this fact has been well publicized). It is important to point out that it would be risky to expect that UNDP will ever again achieve the prominence it has had in regard to mineral exploration in the developing world. This highlights the importance of the new source of financing which the Fund represents.

---

* By the end of 1978 a grand total of $23.8 million had been contributed or committed with the USA joining the original sponsors.

START-UP DIFFICULTIES

I now wish to turn to an extremely anomalous situation. A sizable amount of Fund money is, and will be, available for mineral exploration in a host of countries which are desperately in need of such resources. The terms and conditions for obtaining such assistance are hardly onerous. In fact, they mirror the declarations which have been coming from the Third World for many years now: self-help and union and solidarity so as to present a strong front viz-à-viz the industrial world in order to obtain a fairer share of the distribution of global resources, opportunity, hope and dignity. Nevertheless, for reasons which are difficult to fathom if one thinks in economic terms, the Revolving Fund has achieved only slight acceptance. Although its makeup, its constitution, its aims and objectives were given full approval by the General Assembly and the Governing Council, very few worthwhile requests for assistance have been forthcoming. And negotiation of project agreements has taken place in an unreal atmosphere where it has become painfully evident that the economic and technical factors have been greatly subordinated to political considerations. In fact, a source of disinterested exploration capital which could pump several hundred million dollars into the developing countries during the next 20 years quite possibly is in danger of fading into oblivion. There is no question that there is virtually no limit to the amount of money which will be offered if it can be demonstrated that it will be spent intelligently. To do so, however, will require far more appreciation on the part of potential recipients of what the Fund can do. Even more important at this stage is realization that the Fund can and should represent a major self-help mechanism for that part of the world desperately trying to approach parity with the industrial nations. The revenues which should eventually ensue from the Fund's activities will be earmarked for further work of a similar nature in the Third World. Thus the Fund, if recipients will permit, could be a very practical means of responding in a slight measure to the Third World demands. The Fund alone could, of course, have only a minor impact. But recognition of the fields should give badly needed reorientation of the thrust of foreign aid and transform the pious platitudes concerning mutual assistance between developing countries into something more concrete than shrill rhetoric.

MISCONCEPTIONS

There is a lack of understanding about the Fund, most of which stems from ignoring economic factors.

1. It has been claimed by more than a few that the two percent Replenishment Contribution is excessive, (2) that it will be consider-

ably cheaper to use bilateral assistance. This is highly debatable because:

a. Bilateral aid, even in the form of grants, obligates the government to meet usually about half the bill through counterpart contributions.

b. More often said aid comes in the form of soft loans. A recent example is a 20-year $13 million loan by the World Bank for geological mapping in Indonesia. That the type of work involved may possibly lead to mineral discoveries about 20 years from now is, in my opinion, irrelevant. What really counts is that interest is 8.5 percent per annum.

c. Bilateral assistance travels very much by first class. In the late 1960s at a time when the average United Nations geologist cost the organization $26,000 per year (everything included), because of special circumstances, a subcontract was signed with the United States Geological Survey which provided the services of a few geologists at $62,000 per man-year. And the people involved were middle level only.

d. Bilateral help comes usually from the geological survey or mines bureau of the donor country. By training, philosophy and experience the professionals of these organizations simply are not geared to mine finding. Traditionally their function has been scientific advancement on the higher planes and geological mapping at a more pedestrial level. It has been felt, and rightly so, that these are service organizations providing basic data such as maps by which the mining companies will discover the ore bodies. As a result, insofar as I know, there has been no new mineral deposit of any real significance found, let alone outlined, through bilateral aid during the past 30 years. Or, to put it more mildly, at no time have metal markets been disturbed by waves from bilateral discoveries - not even ripples.

e. About three-quarters of the bilateral budget is spent on the average in the donor country - for supplies, equipment, subcontracts, personnel, administrative and overhead costs.

2. More than once and usually in strong terms it has been suggested that there should be a ceiling in regard to Replenishment Contributions. The La Caridad porphyry copper discovered in Mexico by a United Nations project almost ten years ago is frequently cited as an iniquitous example of the usurious terms the Revolving Fund could exact. When that mine comes into production two or three years hence, it will do so at a scale which would give the Fund revenues of about $4.5 million per annum, or about $65 million over 15 years, if La Caridad had been discovered by the Fund and were obliged to pay the two percent payment which, of course, it is not. As the United Nations investment amounted to only slightly more than 300 thousand dollars, critics have considered a repayment of $65 million as being exorbitant. Certain stark facts of mining life have been slighted in the process.

a. In mineral exploration, the winners usually are defined more cheaply than the losers. The latter, a high percentage of which may be characterized as "teasers," keep dragging the investor on. One more hole, then one more hole, ad infinitum.

b. No major mining company exists which has not had a Roan Antelope or Bingham Canyon to support a considerable number of marginal operations. To put a ceiling on Replenishment Contribution would be to deny to the Fund the sine qua non of any major mining enterprise - the bonanza. To illustrate, with the proposed ceiling of 15 times investment, La Caridad, if obliged to do so, could pay its levy to the Fund in one year. The mine owners could write a check for about $4.5 million while confidently looking forward, at current prices, to receiving gross revenues of about 6,000 million dollars over the next 15 years and at the same time, have considerable assurance that the operating life of the mine could approach a century.

c. Today, investigation of between 500 and 1,000 surface or buried manifestations of mineralization is required to zero in on an ore body of any real importance. (Of course, nobody knows the real figures but these are defensible). With odds of such magnitude, one has to question the capacity of those who conjured up the 15 times investment figure.

3. To illustrate the unrealism of the ceiling referred to above, let us assume that the Fund provided interest-bearing assistance as most foreign aid does. A very simple formula can be developed based on these premises:

   a. According to United Nations experience the average time between the commencement of exploration and the discovery (if any) coming into production is 14 years.

   b. A conservative rate of interest for the years ahead is about ten percent.

   c. Inflation is something which will be with us as long as fiscal authorities reign.

   d. In the developing countries, even with a proven ore body enjoying superior characteristics and location, the minimum risk factor has to be at least two. In fact, in the turbulent world in which we live, two may be considered a very daring figure.

With these parameters, what would the minimum repayment have to be with low-risk investment in feasibility studies? This question can be answered by the formula given below. Its factors are:

C = expenditure of the Fund;

I = average annual interest, which in this instance will be assumed to be 10 percent;

Y = years between commencement of investment and the commencement of replenishment payments;

R = risk factor; and

D = constant dollar factor derived by, let us say, the United States wholesale price index number for the year of commencement of investment, divided by the index number for the year when replenishment payments commence. For ease of illustration the index at the beginning will be given as 100 and that of the year when payment starts will be based on an inflation rate of 5 percent over a period of 14 years. The minimum Replenishment Contribution to achieve revolving status can be calculated as:

$$\frac{[C\ (100) + I \text{ compounded over } Y \text{ years}]\ R}{D}$$

or

$$\frac{(1 \times 2.86)\ 2}{\frac{100}{286}} = 16.4.$$

Thus, a case of dealing with a proven ore body makes a mockery of the ceiling concept when it is realized that the Fund's work will start usually with regional reconnaissance designed to determine whether or not there is reasonable potential for encountering economic mineralization. In fact, as noted earlier, feasibility work centering upon proven ore bodies will not be undertaken during the early years because of the high cost and uncertainty about what would constitute a sound repayment formula. In view of the figures given above, one doubts that governments which want a ceiling of 15 times investment for projects starting essentially from scratch will agree to an equal figure when it is already known that from a technical and geological viewpoint the ore body has been established.

It has been pointed out - in vain - insofar as I know that it will not be in the interest of those concerned more with political than with technical factors, to insist upon a ceiling. The average citizen and the lords of the media know nothing whatsoever about the hazards of mining developments. Coming from environments where doubling your money is considered as a triumph, they will not view repayment of 15 times investment with equanimity. Hopelessly inadequate though it may be, to the man in the street it will seem to be totally exorbitant. But people holding the key to decisions should know better.

CONCLUSIONS

A mechanism for partially solving the lack of exploration money in developing countries could be on the verge of becoming a significant force. Whether or not it will achieve maturity will depend during the next year or two upon objective analysis by governments of the benefits which can accrue to the Third World by encouraging sound and sustained growth of the Revolving Fund. If economic, technical and sociological factors - not political considerations - are paramount, governments will, I believe, eagerly welcome Fund assistance.

7 Are We Running Out of Minerals?*

The publication by the Club of Rome of The Limits to Growth (Meadows et al., 1972) attracted widespread attention. Reviews, comments and critiques were accorded front-page status. On balance, this has to be considered a useful event because it reinforced consciousness of grave threats pressing upon a sorely tried world. It is therefore a pity that the issue which aroused the most concern - nonrenewable resource depletion - is false. I strongly believe that the mineral raw materials base of the world can be considered infinite. Short-term supply problems will exist as long as marketing mechanisms are imperfect, and production of some minerals may cease in the centuries ahead as economic alternatives are weighed; but the totality and mix of minerals required to support a "good life" for a population which the world can feed adequately can be provided at a reasonable cost.

THE EXPONENTIAL

It is only in recent years that exponential patterns of growth have become topical and of concern. First, the population explosion underscored the dangers involved in doubling processes. Later, evidence of the rapidly accelerating nature of so many pollution effects highlighted the drastic need for putting brakes on the endless, mindless

---

* Adapted from a paper prepared for Philippe de Seynes, then Under-Secretary-General of the Department of Economic and Social Affairs of the United Nations, as background material for refuting the main premises of The Limits to Growth (Meadows, et al., 1972) at the 1972 Meeting of the General Assembly of the United Nations.

increase in the dumping of wastes.

In The Limits to Growth the exponential character is dramatized by the legend of the Persian courtier who presented his king with a beautiful chessboard requesting only the seemingly modest reward of a grain of rice in the first square, two for the second, four for the third, and so forth. Well before the sixty-fourth square was reached, the courtier had collared the entire rice supply of Persia. I can give a more dramatic and absurd example: exponential growth of the world's current output of sand, gravel and building stone would result in total extraction of the globe long before the sixty-fourth doubling.

Although the threat to the survival of man posed by rapidly accelerating growth is quite real, valid conclusions will not follow when incompatible exponentials enter the same model. If The Limits to Growth restricted programming to the real and present dangers posed by population and pollution explosions, the contribution would probably not aroused much interest, as these are rather well-plowed fields these days; but the approach would have been scientifically sound. When nonrenewable resources are thrown in, however, the equations lose meaning and validity.

It is pointed out, with regard to potentially arable land, that "within 30 years (or about one population-doubling time) there may be a sudden and serious shortage," (Meadows et al., 1972, p. 51). Although only mentioned in passing, probably far more important as a determinant of how many mouths this planet can feed adequately will be the availability of fresh water at reasonable cost. This is likely to impose harsh limits on exponential growth long before all arable land is under cultivation; similarly with pollution.

In effect, only a few decades are envisioned before crises may be unmanageable. Accordingly, if nothing of major global impact is achieved before the year 2000, it would be more realistic to envision a real decline in the consumption of nonrenewable resources than a continuation of exponential growth. Furthermore, if successful measures are taken, they will tend to be of the braking variety because of the finite nature of food supplies.

The Limits to Growth predicts approximately one doubling time of the key element - population - before disaster strikes if the world continues on its present course. It is therefore quite difficult to reconcile the inclusion of population problems and depletion of nonrenewable resources in the same model. No mineral in commercial use could even commence approaching exhaustion in such a short period. To illustrate, copper is by no means one of the commoner minerals in the earth's crust, yet Lowell (April, 1970) has calculated that "in the first mile of depth there are approximately 3×10^{15} tons of copper metal or about 500 million times the current annual world consumption." He calculates that the concentrated sections with a grade better than 0.25 percent amount to about ten billion tons. It will very likely be economically attractive to mine such material before the end of the present century. This leads directly to as fundamental point which renders invalid the totality of the projections given in The Limits of Growth: with uninterrupted exponential growth - a manifestly impossible happening as long as the human animal requires a stomach - ten doublings would be needed to consume ten billion tons of copper. At current rates of global development, it is most unlikely that copper needs will double as much as twice between today and the year of crisis envisioned by the Club of Rome if the population explosion is maintained. Thus, by taking copper as representative of nonrenewable resources, an extremely conservative approach, it can be seen that entry of minerals into the calculations makes no sense. There is simply no relevance in modeling for happenings 30 years hence with elements which are claimed would bear on the problem only a hundred years thereafter. In fact, as will be demonstrated later, depletion of mineral supplies cannot conceivably ever constitute a determinant of the survival of man.

THE NONRENEWABLE LABEL

The use of the term "nonrenewable" is misleading and in many instances incorrect. Matter cannot be destroyed. It can only be transformed and, in the process, be so scattered as to support the nonrenewable concept. Actually, nature has provided a closed system which, although full of leaks, can and must be made vastly more effective.

In regard to most of today's principal energy sources the leaks are total. Hydrocarbons, once used, are lost forever. (One hopes that the same will apply with fissionable materials!) On the opposite side are the nondissipative character of water flowing into a penstock, the regenerative nature of geothermal energy, and the infinite impact of solar rays.

Elsewhere, many of the leaks can be at least partially plugged. Nature itself is currently doing so on an enormous scale which, in the future, will greatly increase stocks of available minerals. Within the vast reaches of the oceans there is constant precipitation which has, as the principal end product, the deep-sea nodules. That enormous mineral potential does not constitute a resource of only the distant future. The required technology is in the process of development and will probably become operational within a decade.

Particularly with metals, the Law of Conservation of Matter will become increasingly dominant at perhaps exponential rates in the very near future. Recycling for protection of the environment is on the verge of exerting significant force in regard to the reuse of minerals, a process which, on purely economic and technical grounds, has long been important in provision of raw materials. Recovery of secondary metal in the past has been purely an economic matter. Steel mills use more scrap or less scrap depending upon price. The nonferrous metals contained within a piece of abandoned machinery are recovered only if they can be separated one from another profitably. Future forces will end a great deal of such indulgence but not, at least initially, for reasons of conservation. Communities will no longer have the space to pile up garbage nor will public opinion readily accept this solution.

To use the terminology of The Limits to Growth, negative feedback loops are increasingly evident in regard to depletion of mineral resources. Manufacturers are under pressure to develop biodegradable containers and to be constantly on the lookout for easily disposable materials to substitute for those that are relatively indestructible. Increasingly, aesthetics are gaining in respect to sheer materialism, which should further brake pollution and unnecessary wastage of minerals.

MINERAL RESERVES

The treatment given in The Limits to Growth to mineral reserves is both simplistic and dangerously unsophisticated. Any discussion of such a basic and complex element should be carefully qualified in a book written apparently for almost everybody except the earth scientist. It should have been emphasized that:

1. A reserve or resource figure is no more than an intelligent guess based on quite limited data applying only to a brief moment of time.
2. Reserve figures are usually only slightly related to total quantities which may become available in time.
3. Definition of what constitutes a reserve is extremely dynamic and, in a global sense, changes in reserve estimates are almost always upward - sometimes drastically (see Table 7.1).
4. A type of giant exponential is frequently involved. If the grade of a mineral which can be exploited is halved, that rarely means simply a doubling of reserves. Instead, they tend to increase by a factor of ten, a hundred or, to all intents and purposes, infinity. To illustrate, the grade of bauxite which is mined today averages about 50 percent Al_2O_3. World reserves are given as 5.8 billion metric tons and potential ore as 9.6 billion (U.S. Department of the Interior, 1970, p. 446). Eight cubic kilometers of anorthosite, a common igneous rock grading about 25 percent Al_2O_3, contain as much aluminum as the total reserves and potential cited above. About one quarter of a million square kilometers of eastern Canada alone is underlain by anorthosite.

The section entitled "Nonrenewable Resources" in The Limits to Growth opens with a quotation which must have been included only to enhance the climate of fear.

> Even taking into account such economic factors as increased prices with decreasing availability, it would appear at present that the quantities of platinum, gold, zinc and lead are not sufficient to meet

TABLE 7.1. Dynamics of Mineral Reserve Estimates

Mineral		Published Reserves (in thousands of metric tons) Late 1940s (a)	Late 1960s (b)	Percent Change
Iron ore:	actual	19,000,000	251,000,000 (c)	+ 1221
	potential	57,000,000	531,000,000 (c)	+ 832
Manganese ore:	contained (mn)	500,000	635,000 (d)	+ 27
Chromite		100,000	775,000 (e)	+ 675
Tungsten		1,903	1,328 (d)	- 30
Copper		100,000	279,000 (d)	+ 179
Lead		40,000	86,000 (d)	+ 115
Zinc		70,000	113,000 (d)	+ 61
Tin		6,000	6,600 (d)	+ 10
Bauxite		1,400,000	5,300,000 (d)	+ 279
Potash (K_2O)		5,000,000	118,000,000 (d)	+ 2360
Phosphate rock		26,000,000	19,782,000 (d) (f)	- 24
			or 1,178,000,000 (g)	+ 4430
Petroleum:	proved barrels	75,000,000	455,000,000 (d) (h)	+ 507

(a) Based on United Nations (1951), p. 2, Table 1.

(b) Deep-sea nodules not included.

(c) Carman, ed. (1970).

(d) U.S. Department of the Interior (1970).

(e) Meadows, et al. (1972), p. 61.

(f) Phosphorus content.

(g) Emigh (1972).

(h) Not including the petroleum content of oil shales and tar sands which would far more than equal the reserves given here for the late 1960s.

demand. At the present rate of expansion, silver, tin and uranium may be in short supply, even at higher prices, by the turn of the century. By the year 2050, several more minerals may be exhausted if the current rate of consumption continues.

Despite spectacular recent discoveries, there are only a limited number of places left to search for most minerals. Geologists disagree about the prospects for finding large, new, rich ore deposits. Reliance on such discoveries would seem unwise in the long term (Council on Environmental Quality, 1970, p. 158).

Fortunately, the statement is wrong on almost every count. The first sentence infers that a crisis has been reached with lead and zinc. The latest reserves estimates of the United States Bureau of Mines are equivalent, at 1969 rates of global consumption, to 23 years' supply of lead and 22 years' supply of zinc, so that one finds it difficult to understand the cause for alarm, particularly when one considers the growth patterns shown in Table 7.1.

The Bureau of Mines gives a known reserve of platinum in the world of more than 200 million ounces and a production in 1968 of only 1.465 million. But what is not mentioned in the quotation above and perhaps not realized is that platinum has a somewhat inelastic response to pronounced changes in demand. For one of the major world suppliers, the International Nickel Company of Canada, the platinum group of metals are byproducts of nickel-copper output. This group could come under considerable pressure by the market created by new automobile emission standards. However, if demand pressure becomes extreme, it is not difficult to visualize alternative solutions. Finally, as for gold, that metal has apparently been pushed aside from its traditional role in international monetary affairs and the free-market price has doubled recently. In any case, as shown in Fig. 7.1, an industry which for decades increased output despite a fixed selling price, hence a declining real price, should continue to expand now that the shackles have been removed. A world suffering from lack of gold is hard to imagine.

The world's known tin resources - reserves and potential ore - are equal to over 70 years' production at current rates. In the United States, about one third of the metal's consumption is in a form which may be outlawed in the near future - the tin can. Regarding silver, in 1968, almost half the consumption in that country was for silver plating, jewelry and coinage. The world can conserve the metal for more useful purposes; indeed, in regard to coinage, it has already largely done so. Finally, at a price which would have no great impact upon the operating costs of reactors, uranium can be obtained in virtually unlimited amounts from oil shales and sea water. The cost per pound of U_3O_8 would be between 15 and 30 dollars (1) which is not excessive compared to the price paid for Canadian uranium by the United States Atomic Energy Commission ten to fifteen years ago - approximately ten dollars per pound.

To still alarm, it may be advisable to refer to Table 7.1 again. For almost every mineral, reserves have increased, in some instances enormously, during the past 20 years despite exponential growth in

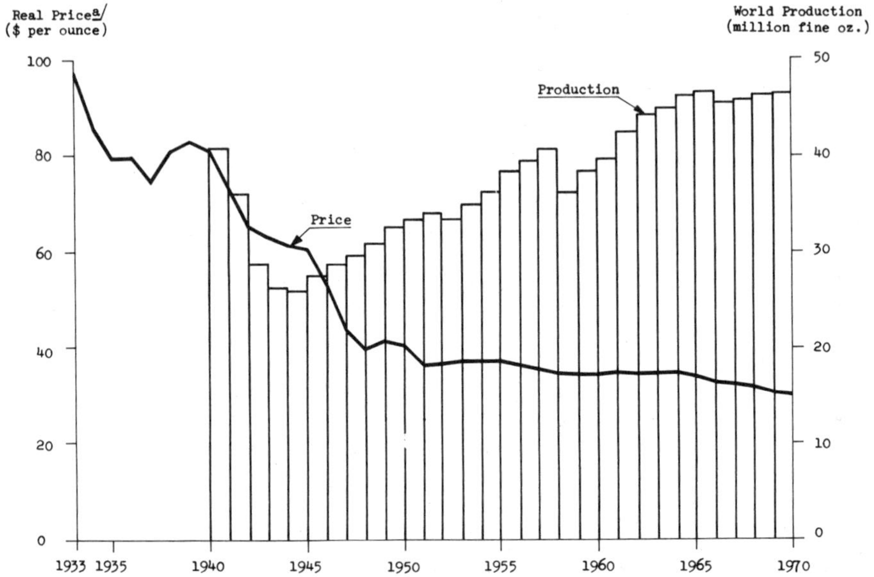

Fig. 7.1. Gold: Production and prices

(a) Deflated by U.S. wholesale price index, 1957-59 = 100.

---

consumption. For only one mineral, tungsten, was there a decline, and that was almost certainly a function of depressed prices. Forward buying contracts negotiated because of the Korean War resulted in a relatively steady and attractive price, averaging $61.64 per short ton in the United States over the period 1952-1956. During the next 13 years the price averaged only $26.92. In fact, tungsten provides an excellent example of interruption of the exponential. Production still has not reached the levels achieved during and immediately after the Korean War.

The statement that "there are only a limited number of places left to search for minerals" is quite astounding. Even if one thinks in only two dimensions, gross error is apparent. Large parts of the earth's surface have been prospected only slightly or not at all - Amazonia, humid tropical belts of heavy laterization; major extensions of the Cordillera and other parts of the circum-Pacific tectonic belt; and, finally, the Arctic and Antarctic. And the vertical dimension has been so slightly probed that only a very small part - a fraction of a percent - of the accessible earth's crust has been even cursorily explored for minerals other than hydrocarbons.

One of the challenges of the age is to develop better tools for investigating the vertical component at a supportable cost. Geophysics will have to become far more definitive, producing at least approximations quantitatively and qualitatively; geochemistry more diagnostic both mineralogically and spatially; and drilling more rapid, easier and cheaper. Above all, the refinement of geological theories and the development of new concepts offer the main hope for discoveries which will meet all foreseeable needs for a world in sensible equilibrium.

RELEVANCE OF THE MINERALS MIX

In The Limits to Growth, reference is made to the fact that the price of mercury "has gone up 500 percent in the last 20 years". That was a most unfortunate example because the price, which reached an all-time high of $725 per flask in July 1965, had dropped to $175 by May 1972, and the likelihood is strong that mercury will sell at relatively low figures for a long time. Recycling has become mandatory with many industrial processes. No longer can wastes with an appreciable mercury content be dumped heedlessly. Thus, in the United States, mercury is being replaced by asbestos in the chlorine-making industry which normally accounts for one-quarter of mercury sales. Another heavy impact upon the mercury market in the United States was made by the prohibition of the metal in paints, an outlet accounting usually for about 15 percent of sales.

Subsequently, the authors of The Limits to Growth say that "the price of lead has increased 300 percent in the last 30 years". At this stage it is not important that ignorance or design is clearly revealed in that no distinction is made between quoted prices and real prices. (I will show later that the real price of the major bulk metals, including lead, has either declined or risen only slightly during the twentieth century.) What I deplore is that individual minerals such as mercury and lead are singled out to prove points without apparent realization of the totality of the minerals mix. A mineral should never be viewed in isolation but should be seen instead as part of a closely interconnected family. Within the group, there will be few functions that can be discharged by only one member. When mercury is banned as a paint base, titanium and zinc take up the slack with a minimum of disruption. If the price of lead really had gone up 300 percent in the last 30 years, most of lead's traditional markets would have been invaded to a greater or lesser degree. When copper prices soar, substitution by aluminium takes place. (2) New minerals enter the picture and, in the process, displace others from traditional markets. Discovery of new uses is a constant process, with an attendant growth in interchangeability. The marketplace will continue to be a balancing force in that it will exert steady pressure toward replacement of relatively costly minerals by those which are cheaper. Scant supplies will lead to the development of synthetics, e.g., artificial cryolite from hydrofluoric acid, quartz crystals from silica and, quite possibly in the not-too-distant future, synthetic diamonds at a reasonable price from carbon.

In effect, there is such constant competition within the family that an exploitive or "make or break" position for any mineral is unlikely except in the short term. There have been two outstanding cases during this century of monopolistic control denying minerals to certain markets. The natural nitrates of Chile were not available to Germany during World War I, so Haber synthesized nitrates from the atmosphere and, in the process, virtually destroyed the Chilean industry. The embargo on the exports of helium from the United States was solved by substituting nitrogen or argon, or by fixing atmospheric helium.

MINERAL PRICES

It can be advanced and proved that during the twentieth century the real prices of minerals have declined on the average. Not only have costs been maintained, but known reserves have increased vastly during the period.

Figures 7.2 to 7.6 are based on United States experience. One can think of no other nation that has so typified exponential growth during the past 70-odd years. As in the case of gold cited earlier, all prices were deflated using the United States Wholesale Price Index, with 1957-1959 equalling 100. The minerals were selected on the basis of the following criteria: (a) availability of reliable statistics (this ruled out nonmetallics because of the difficulty of obtaining historical data and correct prices); (b) large scale output; and (c) relative importance.

The Figures are eloquent testimony to the richness of this Earth and the technical ingenuity of Man. To take copper first, for over 70 years, during which the grade of ore decreased by a factor of about four and output soared, the real price of copper declined. Figure 7.2 also attests to the success of substitution. In 1900, world production of aluminium was a few thousand tons; in 1969 it was over ten million tons, several million more than copper. A considerable part of this growth has been achieved by invading markets traditionally considered the preserve of copper, e.g., in the electrical industry. The door was opened to a large degree because of the highly volatile nature of copper prices.

In the early years, with one aluminium producer dominant not only in the United States but globally, and the market basically only in the process of development, pricing was quite undisciplined. But for the last 30 years at least, aluminium prices have been remarkably stable (Fig. 7.3).

Over the past 70 years, price trends of lead and zinc have been relatively flat, one dropping slightly and the other rising moderately (Figs. 7.2 and 7.5). As for grades of ore, it was not possible to obtain sufficiently reliable data to plot curves. However, there is no question that they have declined during this century but not to the same degree as copper. Most lead and zinc is extracted by underground methods and

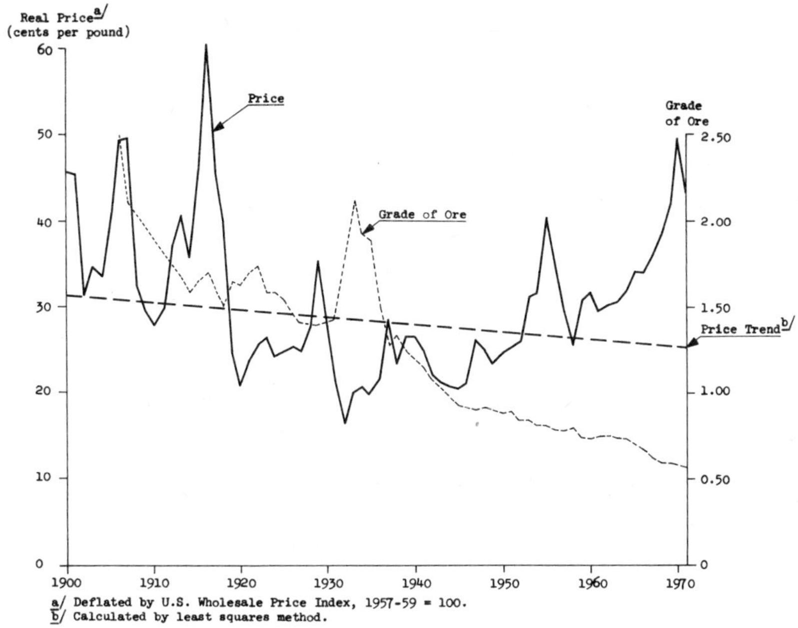

Fig. 7.2. Copper: Prices related to grade of ore.

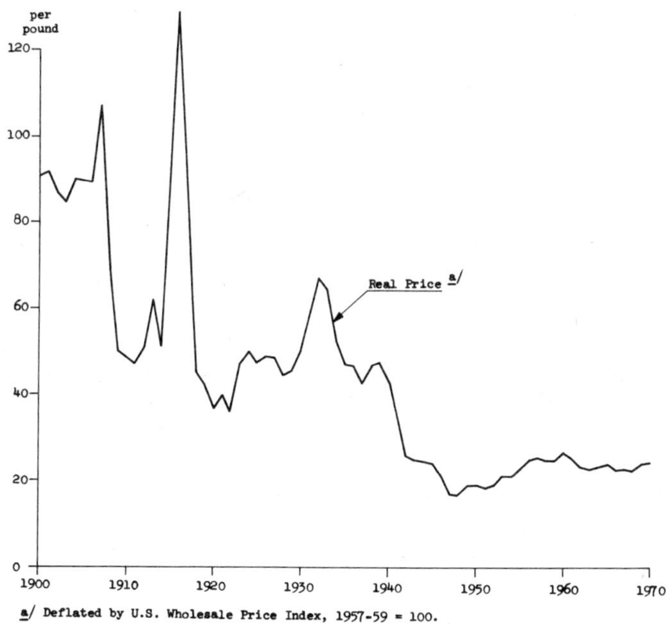

Fig. 7.3. Aluminum: Price Trend, 1900-1970

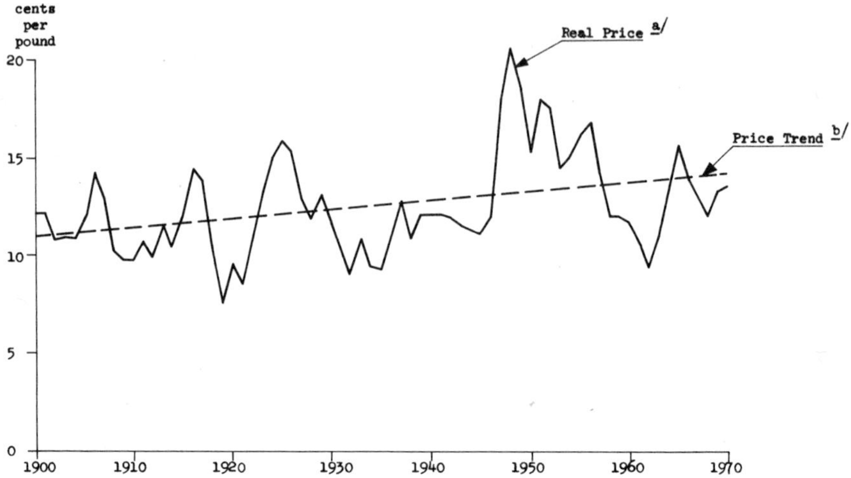

a/ Deflated by U.S. Wholesale Price Index, 1957-59 = 100.
b/ Calculated by least squares method.

Fig. 7.4. Lead: Price trend, 1900-1970

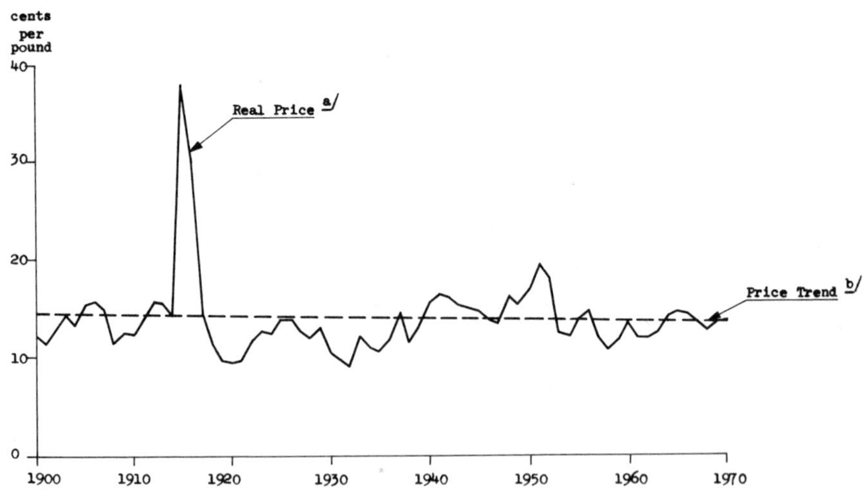

a/ Deflated by U.S. Wholesale Price Index, 1957-59 = 100.
b/ Calculated by least squares method.

Fig. 7.5. Zinc: Price trend, 1900-1970

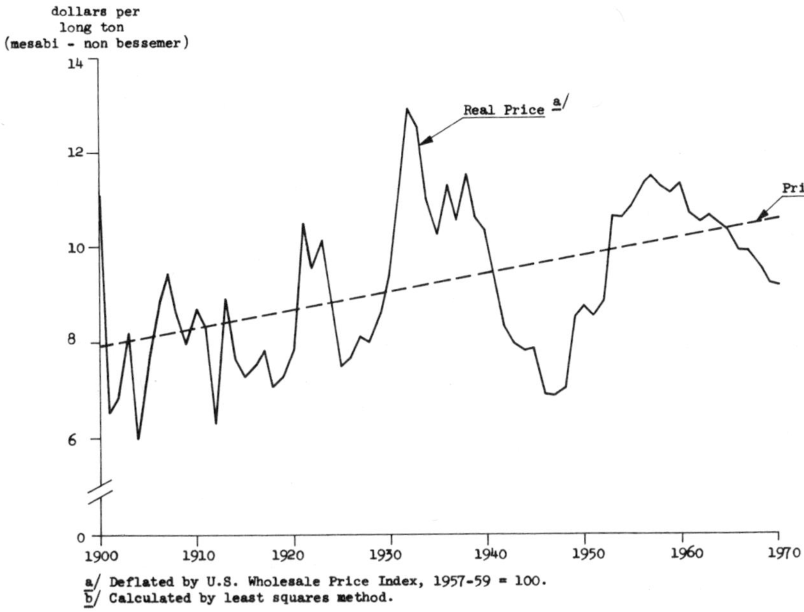

Fig. 7.6. Iron ore: Price trend, 1900-1970

---

is therefore not able to enjoy the economies which accrue to mass mining in open pits. Nevertheless, the industry has enjoyed some significant technological advances, such as the large-scale introduction of trackless mining in the 1950s.

For the purposes of this chapter, the only significant part of the price curve of iron ore is the decline which has been underway more or less steadily since the late 1950s, not only for Mesabi iron ore but also on a global scale (Fig. 7.6). Paradoxically, the decline started at the same time that the concentration of low-grade ores entered the picture in a major way. Actually there is a direct, causal relationship. Twenty years ago there was virtually no interest in the vast low-grade iron ore deposits variously called taconite, banded hematite quartzite (BHQ) or simply "iron formation." With a grade usually about half that required to sell a direct shipping ore, such iron-bearing formations were not even counted as reserves. With increasing emphasis upon pelletized feed for the blast furnace, it was found, particularly where the principal mineral was magnetite, that the grade of ore was a minor element in comparison to amenability to concentration. Thus, with the introduction of a new concept, world reserves of iron ore increased exponentially resulting in a buyer's market which will probably long endure.

TECHNOLOGY

It is obvious that the pricing history given in the various figures is a direct function of technological advances. This poses an extremely important question: Is technology finite?

In achieving, in this century, the production of ever-larger quantities of minerals at lower cost, while greatly increasing known reserves, technological advances have rarely been truly revolutionary. The most momentous event in modern mining, the introduction in 1907 of mass extraction of low-grade (for that day) copper ore at Bingham Canyon, was, in essence, the continuation of a process which started when the first miner was joined by a second. Possibly, the invention of flotation in the first quarter of this century could be viewed as being completely original. Almost all other great steps forward have involved simply the further development of existing concepts, modification of methods, improvement in equipment, refinement of procedures, and the art of management.

The breakthroughs which can be foreseen in the near future are of a similar nature, such as nuclear explosions for in situ leaching, complete automation of processing plants, extraction of deep-sea nodules, great expansion in the production of geothermal energy, and the extraction of minerals from geothermal brines.

I suggest that the advance of technology is not dependent upon luck or the brilliant idea which turns the world upside down. There is no basis for believing that the mental and mechanical acts which led to the rise in productivity of the American miner on a superexponential curve are short-lived phenomena.

CONCLUSIONS

In my view, the concern aroused by the conclusions of The Limits to Growth suggests the need for a plan of action in the mineral field by the United Nations. Let me elaborate:

1. There is obviously considerable need for education and for the collection and dissemination of information on the world's mineral resources which will calm the scene when mistaken doctrine is given publicity. Some years ago, the Resources and Transport Division of the Department of Economic and Social Affairs of the United Nations requested several million dollars to make a detailed inventory of the mineral position of the world. All that was obtained in a functional sense was about $19,000 which resulted in the publication of the Survey of World Iron Ore Resources (Carman, 1970). The sum was insignificant but the findings are such that anybody aware of them could not possibly be concerned about the availability of ample and cheap iron ore into the far-distant future. The same should be done for all of the more important minerals.

2. The United Nations is uniquely qualified as a data collection agency. Accordingly, I suggest that the excellent machinery* which produces the Statistical Yearbook should be expanded so as to provide a world center for the collection, storage and retrieval of data on mineral resources.

3. It is my contention that man can enjoy economic availability of all required mineral resources into a future so distant that time has little meaning. Nevertheless, efforts should not slacken to better evaluate known deposits and seek new ones. One must support strongly the motivation which resulted in The Limits of Growth, namely, that global equilibrium bringing man and nature into harmony is the basic desideratum of the age. To achieve that will require a global strategy. For the world as a whole, the extent to which mineral needs are satisfied will be roughly proportional to the knowledge of the earth's resources. There is thus every justification for the continuation of, and in places the increase in, United Nations involvement in mineral exploration. In many parts of the world, if the United Nations does not undertake the task, nothing will be done, as lease not in the foreseeable future. And it is in the interest of all concerned - the developed as well as developing nations - to encourage disinterested organizations to be in the forefront of the effort to maintain and increase mineral stocks.

4. Exploitation of the vast ocean floor resources is inhibited by lack of rules and regulations. One can therefore see great merit in having the United Nations act as custodian of all mineral resources lying in the ocean basins beyond the continental shelf. Indeed, it is difficult to visualize any other practical means to achieve international agreement for the exploitation of such resources.

5. If only as a function of relative size, the bulk of new mineral resources will be found in the developing countries. Accordingly, the United Nations should assist in determining the amount of processing of minerals that is justifiable in the country of origin. Global consciousness should dictate upgrading and transformation as close to the source of supply as possible if it can be done economically. The developed nations should welcome such "retention of pollution".

U Thant once said: "The truth, the central stupendous truth, about developed economies to-day is that they can have - in anything but the shortest run - the kind and scale of resources they decide to have. ... It is no longer resources which limit decisions. It is the decisions that make the resources. This is the fundamental revolutionary change - perhaps the most revolutionary mankind has ever known" (May 8, 1962).

---

* That of the Statistical Office of the Department of Economic and Social Affairs of the United Nations.

The statement contains a strong element of truth. Unfortunately, it also, by implication, reinforces the fears of those who see a world depleted of minerals. The school of thought which envisions the devastation caused by a five or even ten-fold increase in the consumption of minerals during the next 30 years does so on the basis of a highly erroneous assumption, namely, that the rest of the world will catch up to the opulence enjoyed by the United States. It would be more realistic to expect that the move will be in the opposite direction.

I firmly believe it will not be lack of mineral resources which will put an end to the two cars in every garage dream, to development of outboard motors with the power of a Mack truck, to garbage cans filled to groaning each day with container debris. It will be the fumes in the air and the slick on the sea which will finally force a halt if there is to be meaningful survival. The high tide of sheer materialism may be very close. Then, as the act of possessing abates, a fundamental truth may dawn upon human consciousness: once basic needs for nourishment, physical, mental and aesthetic, are met and shelter is assured, the quality of life will be found to be only superficially related to goods and chattels.

8 Taxation of Mining Enterprises*

INTRODUCTION

An examination of international taxation policy can be approached in different ways, principally in terms of statistics or in terms of issues. Tabulations of types of taxes and rates on a country-by-country basis may appear impressive but actually, they are almost always misleading and even incorrect. Reliable statistics are very hard to obtain. In many countries taxation law is so complicated that it constitutes a jungle in which it is very easy to get lost. To illustrate, as of ten years ago in Chile at least five different tax codes were in existence: (a) small mining; (b) medium mining; (c) gran mineria; (d) a law offering a tax-free period of 20 years to new mining enterprises of small or medium size; and (e) a decree passed in 1959 which granted new mining ventures in certain economically lagging provinces remission of 90 percent of duties and taxes on income until 1972.

In many developing countries, sheer frustration and a desire to try anything which will change the status quo leads to a proliferation of statutes.

Too many enactments are made as a sop to public opinion without the enabling regulations being developed. It is therefore a difficult and frustrating to determine what the law is. A legal opinion may be useful or quite disastrous. For years I ignored a rather formidable billing from

---

* Based substantially on a paper with the same title delivered at the United Nations Inter-Regional Seminar on Mineral Economics, Ankara, Turkey, October 1970 and translated into French by Pierre Legoux for publication in Annales des Mines, Paris, December 1971, under the title "Fiscalite des Enterprises Minieres."

an august firm in Brazil which had been engaged to give interpretations of four relatively simple points of mining law. Subsequent investigations revealed that three of these were highly erroneous. However, legal confusion is by no means a monopoly of developing nations. In the United States, for example, the indispensible corporation men are the tax lawyers and accountants.

Furthermore, most mining statutes are weak in that they have been only imperfectly responsive to the revolution in mineral development of the past 25 years. For these reasons, I shall focus here on selected issues, rather than statistical comparisons, with reference to the fiscal framework within those few countries toward which the largest part of international mining investment has been flowing, and on possible future trends in the developing countries.

FACTORS WHICH CONCERN CAPITAL

Stability

Assuming favorable geology, the most important determinant of where mining investments will take place is fiscal and political stability. Of course there have been many instances where political turmoil has had little or no effect upon the day-to-day conduct of business. But by and large, revolutionary environments, whether of the Right or of the Left, do not create comfortable climates for private capital. Newcomers will be discouraged, and those already locked in will tend to use various hedging devices which will hurt the country in the end, such as high-grading and limiting reinvestment to a dangerous degree by running down plant and equipment and reducing working capital and inventories.

Stability is even more important in attracting investment than taxation rates. To illustrate, regulations are tough in Mexico: taxes are high, and controlling interest in the mining enterprise has to be in the hands of Mexicans, either individual or corporate, or both. Nevertheless, a great deal of international mining capital is flowing into the country largely because the relatively stringent laws are more than counterbalanced by several decades of political stability and by a sound currency. In addition, as mining has lagged well behind general economic development in Mexico, it is a good bet that legislation affecting the industry will be made more liberal rather than more stringent. This is important because it provides a basis for calculating the cash flow of a venture with some degree of confidence. That the grade of ore may have to be considerably higher to ensure a profit than with a similar mine in Canada or in the United States is not a basic issue. What counts is the advantage of having some indication of what could be the most adverse impositions of the government.

Period of Amortization

Rapid return of capital is, of course, a fundamental factor in the equation. Many, if not most, international mining houses will not take on anything when the indicated amortization period is longer than seven years, particularly in new countries. The principal elements which determine the rate of recovery of capital are:

A Tax-Free Period in the Early Years

This may vary from three years, as in Canada at the moment (1970), to being open-ended. The method whereby taxes on income are not imposed until after recovery of investment is frequently better for all concerned than a fixed-term system. The latter creates the temptation to adopt a short-sighted policy of maximizing profits in the early years by concentrating on extraction of the richer parts of ore bodies. Rapid depletion of what, with good mining practice, should be sparingly doled out as a "sweetener" can seriously shorten the life of the mine. Fortunately, with a growing proportion of large new mines, opportunities for high-grading, are quite limited. With low-grade dissemination, sizable zones with a tenor well above the average may not exist. In addition, the massive scale of operation which is becoming prevalent limits the opportunities for selective mining.

Accelerated Depreciation

In Canada, for example, the investment in mine plant and equipment can be written off at up to 30 percent per annum, computed on a diminishing balance basis. Shafts, haulage ways and similar underground workings are treated as an operating expense and thus written off at a rate of 100 percent. Capital cost allowance is not compulsory; hence, any write-off up to the permitted maximum may be taken in the fiscal year concerned. Accordingly, depreciation charges are not entered into the books until termination of the tax-free period. At the other extreme, one can note legislations under which plants and many equipment items have to be amortized over as many as 30 years. This may be logical with a textile mill, but it does not make sense with a business as risky as mining, particularly with medium-sized mines where the reserves may be sufficient for no more than a few years.

Depletion Allowance

For a long time, the concept of depletion was accepted in important mining countries such as Canada and the United States as recognition of the wasting nature of extractive operations. Today, the word "depletion" is very much out of favor. The argument that a company should obtain special concessions because it is consuming its capital, the ore body, wears a little thin after the investment has been amortized ten times over. In addition, and even more important, in recent years it has been held in most parts of the world that subsoil resources belong to the state. Accordingly, the entrepreneur cannot claim as capital something which does not belong to him, and for which he has only a license - often on a fixed term license - to exploit. All in all, because of the controversies and emotions they give rise to, depletion concessions should perhaps not be included in mine taxation codes. However, simply exorcizing the devil does not solve the problem. Mining remains risky and account must be taken of this factor. There are various ways in which this can be done: rapid write-offs, a more lenient tax base than with conventional industries, elimination of duties and sales tax on process equipment, etc.

Simplicity

If a battery of lawyers must constantly work overtime in the planning and implementation of every step leading to a viable mining operation, few investors will be interested. The business of bringing in a mine is tough enough without having to become entangled in a maze of often confused and conflicting laws. Where the rules of the game are not reasonably well defined most people simply decide not to play. Actually, there is no reason why mining statutes and regulations cannot be written with clarity and simplicity. For several decades mining prospered greatly in Canada, and with ever-increasing revenues to the state, under a very brief fiscal code which even a schoolboy could understand. (Unfortunately there was a complete turnabout in Canada after 1971. This is chronicled in Chapter 9.)

In addition to simplicity of legislation, clean cut provisions and regulations are vitally important. A situation can arise where tax rates remain stable or increase only slowly but methods of tax collection change frequently. One year the cost of a new shaft is treated as an operating expense, in the next it is ruled that it has to be amortized over a period of years. The 1967 car imported for the manager was depreciated over three years, the 1970 model over five. Frequent and often arbitrary changes in administrative procedures do not create a climate of mutual trust even when they have no significant impact upon the financial fortunes of the enterprise.

Taxation Not Related to Profits

Capital is, in general, quite wary of environments where significant taxation is imposed whether a profit is realized or not. Thus, export taxes and excessive duties and royalties on gross are not viewed with favor. The export tax is considered to be particularly pernicious. In more than one instance, important segments of a nation's mining industry have been priced out of the market by these. As for duties, there is no question that they are a necessary mechanism in most developing countries for stimulating the growth of industry and conserving foreign exchange. Indeed, in many countries the customs house is the principal generator of public revenue. However, as long as many of the most active mining nations of the world offer partial or total remission of imposts on equipment imported for primary industry, governments concerned with attracting mining capital should cast a critical eye upon customs tariffs. Finally, a royalty on gross can be a decided cause for alarm. This chiefly stems from realization that many taxing authorities have little or no appreciation of how heavy the impact can be. Accordingly, it is very tempting to keep increasing the royalty with almost every new budget by seemingly insignificant amounts - let us say, one or two percent, à la New York State's sales tax. Unfortunately, what many governments do not appreciate is the rather small profit margin of most mining operations. Even a royalty of two or three percent can have a formidable effect upon a company's fortunes.

I shall refer later in this chapter to another adverse feature of royalties from the viewpoint of certain investors. I should note here, however, that there is another side to the coin. The various revenue producers described above have the great advantage of ensuring a steady flow of income throughout the year. Even royalties are often paid in kind, and therefore concentrates continually move into government bins, whereas the income tax is usually paid only once a year. To resolve this time lag, one method which warrants consideration is to assess export taxes or royalties as an advance payment of income tax. However, should an enterprise suffer a loss in any given year, export and royalty taxation should cease until profitability is restored.

Carryover of Losses

Most countries where the mining sector is dynamic permit the losses of preceding years to be subtracted from profits of the current year in whole or in part before computing taxable income.

Need for Moderation

The majority of international mining capital is being invested today in four countries - Australia, Canada, South Africa and the United States. This is not primarily a function of geology. Per unit of area, probably the richest copper province of the world is the Cordillera. In British Columbia, Canada, a very small part of the whole, the first porphry copper mine came into production in 1962. Since then the record with the larger new mines either completed or pending is as shown in Table. 8.1.

TABLE 8.1 Investment in Major Metallic Minus in British Columbia: 1962-1969

Mineral	Number of Mines	Combined Concentrator Capacity (tons/day)	Investment (Canadian dollars)
Copper	8	60,000	305,000,000
Copper-molybdenum	6	175,000	530,000,000
Molybdenum	3	35,000	70,000,000

During the same period, far to the south in the Andean countries, total investment has been far less, even where concerns with an already major vested interest have been susceptible to pressure for expansion. And yet, the Andean copper deposits are generally of far higher grade and located closer to ports. The major Andean copper mines operate on 1.2 to 2 percent copper (Cu) with byproduct molybdenum (Mo) credits in some cases. This is certainly far superior to almost all of the British Columbia mines. Lornex, with 0.43 percent Cu and 0.014 percent Mo, and Brenda, with 0.19 percent Cu and 0.08 percent Mo illustrate the lower limits but are not much below the general average.

India and Brazil have long been noted for having enormous deposits - many billions of tons - of high-grade iron ore. The locations of these have been no better or no worse on the average than the deposits of Australia, Canada and Liberia, all of which have been actively encouraging development of the iron ore industry. The trends since 1961 (see Table 8.2) when Australia lifted the embardo on exports of iron ore, are particularly revealing.

TABLE 8.2. Iron Ore Exports (in thousand metric tons)

Country	Brazil (a)	India	Australia	Canada	Liberia
1961	6,078	9,906	-	15,124	2,840
1962	7,649	9,086	-	22,019	3,806
1963	8,268	9,880	-	24,266	6,465
1964	9,730	10,482	85	30,425	12,237
1965	12,731	11,264	1,557	30,676	15,347
1966	12,910	13,569	1,970	30,693	16,571
1967	14,279	13,575	8,895	31,481	17,191
1968	15,050	15,650	16,338	36,032	21,668
1969	N.A. (d)	16,670	20,072 (c)	N.A. (d)	22,379

(a) A decided change toward policy favoring free enterprise catapulted Brazil in a very few years into first place as an iron ore exporter. In 1977 the total was 73 million tons. Today (1978) that country sets world prices and usually sets them lower.

(b) Including Goa.

(c) For fiscal year July 1, 1968 - June 30, 1969.

(d) N.A.: Not available.

It is significant that the largest part of iron ore mined in Canada has a grade less than half of that being extracted in India and Brazil. Also noteworthy is that whereas Australia had known reserves of only 500 million tons in 1961, the figure increased to over 16 billion six or seven years after the export embargo was lifted, and it is now evident that the potential far transcends known reserves.

The above highlights the importance of moderation, i.e., of having a cold-blooded appreciation of the value of the cards in one's hands. Usually, the developing country does not hold many aces when it comes to dealing with new sources of capital. If the terms and conditions are considered onerous when related to risk, the investor will simply move elsewhere to sit in on a better deal from his viewpoint. This may seem to be no more than a statement of the obvious but unfortunately, in too many places and too frequently, it is not realized; governments very often simply overestimate their bargaining power.

The country in the weakest position is one where the mining industry is quite immature. In such cases, unless a bonanza of major magnitude has been proved, it will almost always be sound policy in the beginning to open the door wide with a minimum of terms and conditions if private capital is desired. At a later date, when the stake of the investor has become large, arrangements can be made which will serve national interests better. In the final analysis, governments are the controlling power. I am not suggesting that power should be used in an inequitable fashion. I am, however, proposing that once the enterprise has amortized its investment, the government can legitimately and more easily claim a larger share of the cake. All governments do. But in the beginning, exigent demands will rarely elicit any meaningful response.

Repatriation of Capital and Profits

The potential investor has to examine very carefully the past record and present policy in regard to repatriation of earnings and capital. Impediments to the free flow of capital are not conducive to investment.

Multiple Exchange Rates

Multiple exchange rates are always discriminatory and often aimed at the exporter by forcing him to surrender to the central bank foreign exchange earnings at an artificially low rate. I can recall that in Bolivia, before nationalization of the mines, the dollar was selling on the black market for about 200 bolivianos. The black market is usually a better indicator of value than rates fixed by government edict. Therefore, very heavy taxation was being imposed when the central bank paid exporters only 40 bolivianos for the dollar.

INDICATED TRENDS IN DEVELOPING COUNTRIES

In the discussion which follows, the word "taxation" is rather loosely defined. It embraces any mechanism whereby a government receives a share of the earnings or of the equity.

Tax Rates

In the long run, it is almost certain that the developing countries will obtain an increasing part of the proceeds from successful mining operations in their territories. The pressures in this direction stem from:

1. The fact that most countries have established their sovereignty over subsoil resources, excluding, in many places, only such materials as sand and gravel.
2. The growing tendency to insist that a controlling interest, usually 51 percent of the equity, be in the hands of nationals. Of course, this can be circumvented by setting up dummy shareholders, but in the long run this is a dangerous procedure.
3. The need of such countries as Japan, the Federal German Republic and the United Kingdom to ensure continuing supplies of minerals in which they are lacking. In some cases, these countries will take on what are virtually break-even propositions in order to adequately supply the rod and brass mills at home. With respect to mineral raw materials, these are essentially have-not nations and, therefore, not blessed with many options. Even a country with such enormous mineral production as the United States is a net importer of mineral commodities.

Recently, a Japanese firm, Overseas Mineral Resources Development Company, negotiated a concession in Ecuador covering a potential porphyry copper deposit discovered by a United Nations-Ecuadorian team. The terms may portend the future: the royalty on gross output was 7.5 percent on a grade of up to 0.9 percent Cu and Mo equivalent combined, rising on a more or less straight line curve to 18.5 percent for material grading 2.5 percent or higher; and the income tax was 55 percent of net profit. If this becomes the norm in the years ahead, one can visualize that only fully integrated concerns will be able to undertake foreign ventures in many countries. They will do so on the basis that relatively small profits, at least from production of minerals, will be more than counterbalanced by the advantage of securing supplies which permits gaining added value from transformation into semi- or fully fabricated products at home.

Negotiated Tax Structures

Particularly in countries where the industry is small, most truly significant mining development in the future will probably require special legislation. Relatively few countries have a mining code sufficiently well designed to be able to cover the entry of perhaps hundreds of millions of dollars for a single project. Each case has to be distinctive because of a host of factors which differ widely from place to place: characteristics of existing and needed infrastructure, availability and quality of labor, location and type of deposit, existence or lack of supplies and materials within the country, tax regime of the nation of origin of the investors, etc.

The last point is particularly important. Blanket regulations may contain concessions which lose all meaning because of the laws of the country from which the development capital originates. The United States holds that unless there is a bilateral treaty to the contrary, profits of foreign-based subsidiaries are taxable at home. Thus, a tax-free period has no significance for the company unless the rates abroad are higher than in the United States. On the other hand, for the government such a concession simply means enrichment of the treasury in Washington at the expense of the country itself. Another example is the royalty. It is treated in the United States as an operating cost and therefore cannot serve as a deduction in its entirety from taxes owed to the federal government. Hence, the ideal treaty for United States companies would involve no royalties, income tax rate no greater than at home, and accelerated depreciation. Companies from countries which do not exact tribute on foreign-based operations would prefer the reverse, a tax-free period and deferment of write-offs until its expiration.

Fixed-Term Exploitation Licenses

It is becoming increasingly common not to grant mineral rights in perpetuity. An acceptable agreement today in many parts of the world is to grant an exploitation right for about 30 years with the possibility of negotiating an extension. Thirty years may appear to be a very long time. In actual fact, some mines have lasted for hundreds of years and there are deposits, particularly of nonmetallics, which are virtually inexhaustible. Hence, with properties of major size, if terms and conditions are honored, the end product of this policy is likely to be: (1) the nation will eventually become the full owner, or (2) the company will scale operations to achieve the maximum output consistent with the period of the license. In either case, the country stands to benefit.

Tax Incentives for Local Processing

At any given time it is difficult to determine exactly what is the optimum form in which a mineral product should be exported. Exporting nations desire invariably to process it as much as possible before export, whereas it is usually in the interest of the importers, the industrialized countries, to obtain their needs in the most primitive form, concentrates or even occasionally hand-cobbed ore. There is a basic conflict here and it must be approached objectively. Where there is only an insignificant local market, lack of trained personnel, high power costs, etc. - common disadvantages in many developing areas - there is nothing to be gained and a great deal to be lost by trying to force uneconomic processing. Then, there is the matter of scale. No legislative fiat is going to lead to a full product line of reasonably priced steel where the market is only a few hundred thousand tons annually. Similarly, it requires a very large base-metal mine usually to support an economic smelter and refinery.

Despite the cautionary words above, governments and private capital should always be alert to and constantly reappraise opportunities to build up local industry. In the early stages of development of the mining sector, this will usually occur with nonmetallics. These normally can be converted into consumer products with far less investment than metals and have greater potential for creating a significant internal market as, for example, phosphate rock converted to superphosphates, or gypsum to plaster of paris and wallboard.

To initiate such processing, when appropriate, government should give serious consideration to the use of tax incentives. Usually revenue to the state will increase significantly because of added investment and the multiplier effects stemming from increased employment in production and distribution.

Joint Ventures

There is a very good chance that in the years ahead the usual corporate vehicle for mining operations in developing countries will be the joint venture. It has some very attractive features:

1. It increases public revenues, assuming profitable operations.
2. Participation by government tends to dampen the passions which get aroused by the concept of the extraction of nonrenewable resources under the total control of foreigners.
3. It aids in development of concepts of modern management.
4. The sophisticated potential investor will usually welcome partnership with the government as a means of obtaining protection against expropriation or nationalization.

When relative shares are worked out, the following factors will enter the equation:

1. Companies will be reluctant to give more than token equity to the government at the high-risk stage of exploration. The odds against finding an ore body are usually so great that mere title to subsoil resources is not negotiable at a high price.

2. Where work by the government has revealed something of interest, determination of respective shares will tend to be directly proportional to how well defined the deposit is both in the physical and economic sense.

3. In the case that private capital is offered a part in the exploitation of proved tonnage of a grade and character which more or less assures an economic operation, the government will often be in a position to demand a majority interest. Occasionally, this will be acceptable provided that the investor obtains full management powers for at least as long as is required to amortize plant and equipment.

4. Most cases will fall in between the extremes defined by "wildcat" exploration and a proved ore body. Respective equity positions will then be defined in terms of what each party can contribute. At this point, it would be a mistake to overestimate the value of the potential ore body. The path from the point of finding interesting mineralization to a profitable operation is usually full of quicksand and boobytraps. However, let us assume that agreement is reached as to the equity to be obtained by the government on account of the possible ore. The next stage, the risky business of outlining the deposit, should be undertaken entirely by the company. Assuming a successful conclusion of this, the government often should participate in financing up to the production stage and thus obtain additional equity. To what degree this will be done will be a function of a thousand variables. One approach that can be recommended is contribution on the basis of ability to pay: for example, the government meets those costs (chiefly for infrastructure and labor) payable in local currency, and the private concern finances those supplies, equipment and technical skills which have to be paid for in foreign exchange.

CONCLUSIONS

Certainly, no single factor is as important a determinant of whether or not a mining investment will be made as taxation policy. Under any private enterprise system the most important questions is: "How will the gains be divided?" Each party to the contract has to be antagonistic in the sense that a percentage point picked up by one reduces the take of the other.

Under the circumstances, seeking and finding the happy medium, the ideal, will be more or less akin to attempting to find the Holy Grail. But somehow or other accommodations have to be made. Except in centrally planned economies, marriages have to be arranged between private capital and governments to bring new mines on stream. To do so in a world seething with undercurrents, with open confrontation between the haves and the have-nots, will require statesmanship of a rare order these days. I suggest that the tax regime which will work, i.e., be acceptable to foreign concerns, will be one which permits rapid recovery and repatriation of investment, mineral prices permitting. Subsequently, more or less equal sharing of net profits appears to be a reasonable formula, adjusted, of course, for respective equity positions in the cases of joint ventures between private investors and governments.

In the final analysis the state should view a new mining enterprise as a source of taxes only in a secondary sense. The primary interest should rest on such things as infrastructure creation, downstream effects, employment, the enhancement of skills, and sharing of know-how.

9 Evolution of Mining Taxation Policy in Canada*

H.G. Wells once said: "No passion in the world, no love or hate is equal to the passion to alter someone else's draft." The keen insight of that eminent writer notwithstanding, my passion is most readily aroused by the idiocies, not to say inequities, of governments and the social scientists who advise them when confronted by the unknown, a category into which mining in Canada and many other places fits if recent events are an example. In Ottawa and the various provincial capitals, the mining industry has been viewed too often as an enticing wilderness, one into which the politician and public servant bravely and eagerly venture - almost invariably without machete or compass. In this chapter, I shall trace the effects of this ignorance on recent policy and discuss their implications.

THE BEGINNINGS

It all started innocently enough with the British North America Act, the legislative vehicle for the federation which created, in 1867, the Dominion of Canada from a number of disparate colonies which, on the whole, had not much in common except geographical proximity. Deeply concerned by a very recently concluded civil war which had all but torn the United States apart, it was the intent of the framers of union that the central government would fall heir to residuary powers, that the provinces would be strictly circumscribed in regard to rights and authority. Cardwell, British Colonial Secretary, the key figure in

---

* Presented in substantially the same form at the Curso "Roso de Luna," Escuela Tecnica Superior de Ingenieros de Minas, Madrid, Spain, May 23-27, 1977.

obtaining ratification by Whitehall of the Articles of Confederation, wrote: "My principal fear is that the selfish interests of the men of the smaller provinces should lead them to insist on federation as against union...on a policy which would run British North America upon the rock on which the Union has gone to pieces" (Creighton, 1956, p. 389). Cardwell believed that a federation might be constructed with sufficient strength and centralization to, in effect, amount to a legislative union. "Of course," he wrote, "it signifies very little what name is employed. What we wish is a central and strong government as distinguished from a number of small states united by a feeble bond" (Creighton, 1956, p. 389). Ironically and sadly, it appears that Cardwell's fear has become reality.

The British North America Act, like all such legislation, was concerned with who would do what, and who in any particular circumstance was the boss. Some of the most important provisions of that document deal with the division of powers between the federal and provincial governments. "These powers cover the whole area of government in Canada and each level is sovereign with respect to the power it exercises. Hence provincial governments when acting within their jurisdiction as set out in the British North America Act are as sovereign as the federal government..." (Canada Yearbook, 1975, p. 36). Federal power, as one would expect, covered common affairs - defense, currency, weights and measures, criminal law, etc. Provincial powers seemed limited to matters of local concern - education, the management and sale of land and timber, direct taxation within the province and the like. No one gave any thought to a future which would spotlight mining as one of the most important prime movers of the entire economy.

In the original conception direct taxation was envisioned as being very simple indeed: ten shillings on each barrel of rum imported for consumption within the province, sixpence to cross on a ferry, that sort of thing. A half century was to pass before the inconceivable took place, namely, taxation of corporate profits. When it was introduced during World War I at a rate of two percent, the industrialists, no doubt, would have felt that the ghost of Karl Marx was stalking the land had they ever heard of Das Kapital. Later came social welfare, a gentle revolution which had to upset all calculations and foresight of the Fathers of Confederation. In 1867, the responsibility of the state toward the afflicted was essentially limited to those horrors of the Victorian Age called the "poorhouse," "workhouse" and "orphanage." The Fathers could not have envisioned unemployment insurance, extremely costly and almost free medical care, maintenance of the aged, and subsidies for childbearing - in sum, a changing fabric of social responsibility which has placed burdens upon the provincial governments beyond what can be met by such revenue producers as gasoline, liquor and sales taxes.

The drafters of the British North America Act aimed at a division of powers which was clear-cut and which would stand the test of time. The intent was sound, statesmanlike and, for the day, broad gauged. Unfortunately, none could possibly foresee the turmoil and complexities which the twentieth century was to inflict upon long-established theory and practice. It has been evident for a long time that the assigned powers were not mutually exclusive and frequently overlapped.

PROVINCIAL-FEDERAL CONFLICT

It would be a big mistake to view Canada as a nation. In its stretch from ocean to ocean, it cuts across several regions, each with quite distinctive levels of natural wealth and development; it has to hop language barriers; and there is the strong and constant pull southward which finds the man from Vancouver having far more in common with his neighbor in Seattle than with his countryman in Halifax, 3,000 miles away. Geography, politics and Canada's constitution have created something of a monster. Instead of a nation with firm centralized control and direction, there are, in effect, eleven almost sovereign bodies, each inward-looking and each wanting a larger share of the pie.

Until well after World War II, the provinces, although undisputed owners of subsoil resources, extracted only minor taxes from the mining industry. The federal government took the big bite and, under tax-sharing arrangements, kept the provinces more or less satisfied. But beginning in the Fifties, a number of developments have led many provincial governments to view the mining companies as an ideal target for extorting the revenues required to support, among other things, the welfare programs deemed essential to win office and stay there.

1. From the United Nations came the declaration of the permanent sovereignty of nations over their natural resources, which has been widely publicized and often hailed. To make a resounding proclamation in regard to something which has been self-evident since the first national boundary was manned by troops cannot be dismissed as an empty gesture. Its effects have been serious. Too much emotion has been aroused; too much attention has been focused upon an industry which has a vital role to play in the troubled years that lie ahead; and, in the process, a lot of ammunition has been provided to the demagogues seeking catch phrases such as "protection of sovereignty," "guarding the national patrimony," etc. In recent years, Canada, a country where political battles have traditionally been moderate in tone, has produced its share of people who can raise shrill and impassioned voices over an issue which should not be an issue.

2. The federal government opened the door to change by appointing a Royal Commission on Taxation in 1962 to look into taxation in general, including, of course, those imposed on mining enterprises. I shall describe the findings of the commission in the next section. Broadly speaking, with regard to mining, these amounted to a recommendation to tear apart a tax system which, for decades, had been universally recognized as the best in the

world for stimulating mineral development while ensuring a fair share of revenue for the provincial and federal governments. The end result was that the mine taxation regime was radically changed and in the process it was revealed to provincial authorities that there existed the most attractive of all targets from a politician's viewpoint - a beleagered industry, held in general disfavor, from which greatly increased revenue could be exacted, both monetarily and in terms of political capital.

3. The mining industry, with a number of expedient and callous acts at precisely the wrong time, seemed determined to commit suicide as I shall explain later in this chapter.

When the provincial governments started to cast covetous eyes upon mining revenues, the industry, to its great distress, found itself a victim of the lack of clear-cut distinction between the various powers to tax. What has taken place in recent years has been a savage, destructive and, at times, insane tax war between federal and most provincial governments in regard to resource industries. Inevitably, the war escalates. As one province exacts greater tribute, the leaders of the rest find themselves under pressure to follow suit or to even outdo the others. Waddingham (January, 1976) has charged that increased federal and provincial taxes directed against the mining industry since 1972 have led to an "...incredible taxation nightmare.... Since taxes vary from jurisdiction to jurisdiction within Canada, we now have inter-provincial rivalry emerging on the basis of taxation. Average return on the invested capital for the total mining industry over the past five years has been less than the returns from manufacturing industries."

As the internecine battle has mounted in ferocity, federal authorities have tried to counter the measures by which the provinces have sought to maximize their take, thereby eroding the revenues accruing to the central government. No longer does the federal government allow the deduction of royalties imposed by the provinces from federal income taxes, nor their treatment as a cost of operation: double taxation with a vengeance!

Much that is happening these days is simply not understandable. Accordingly, let us go back to the beginning, when the cash flow and the allure of the Canadian mining industry were changed in an extremely adverse manner.

THE CARTER COMMISSION

The Royal Commission on Taxation which I mentioned earlier and which has been generally called the "Carter Commission," after K.L. Carter, its chairman, had a formidable task, namely, to critically examine the terms and conditions which had been the basic pillars of a widely acclaimed taxation regime of long standing. With respect to new mines, there were:

1. Exemption from taxation during the first three years of production.
2. Automatic depletion allowance which amounted to one-third of net income before taxes.
3. Write-offs commencing after the expiration of the tax-free period at the rate of 30 percent per annum on the declining balance. (In effect, this amounted usually to almost full amortization of investment from a bookkeeping viewpoint in seven years, during which tax liabilities were minimal, to say the least.)
4. Deduction of 100 percent of exploration and development expenses in the year incurred against income from any source with an indefinite carry-forward of any costs not written off.

When the Carter Commission's report was tabled in 1967, it was evident that it bore hostility to the mining industry, having found that the risks involved in mining are no greater than in other industries. The commission was clearly in error in this respect. The run-of-mill of business enterprises are set up with market surveys and process analyses which may cost a few hundred thousand dollars. In contrast, the expenditures involved in finding a major mine in Canada today will be in the neighborhood of 30 million Canadian dollars on the average. (1) Furthermore, with the vast majority of manufacturing concerns, pricing will be a relatively local matter: as new wage rates are granted, as suppliers raise their prices, adjustments are smoothly made to pass on the increases to the consumer. It is not quite so easy in the minerals market. A certain mine in British Columbia may be running deeply in the red, but it will have no chance whatsoever to obtain relief if there is abundant low-cost Chilean copper seeking a market. Very few nonspecialists realize that the mining industry, which has done a good job in providing the raw materials needed by man, has done so at prices which, measured in constant dollars, have generally declined during this century despite ever-decreasing grades of ores.

In addition, the Carter Commission questioned the need for the incentives provided by the three-year tax exemption period and the depletion allowances on the baffling grounds that these benefited mainly the large mining companies which had little need for them. It is more likely that the carrot which the tax-exempt period represented was of more value to the small promoters, the progenitors of an amazingly high percentage of Canadian mines in other days. These true pioneer entrepreneurs did not have capital; therefore they had to have selling points, such as exemption from taxation, when they talked to the bankers or touted stock issues. Nevertheless, although one questions the reasoning and many of the recommendations of the commission, one has to agree that it was correct in recommending an end to tax-free rides and automatic depletion allowances which were in no way related to the life of the mines or to efforts to replace wasting assets.

No doubt, the mining industry of Canada had done much to dig its own grave, even before the Carter Commission. Over the years, pressures were usually heavy to maximize earnings during the tax-exempt years. But it is important to note that two of the most blatant instances of cold-blooded high-grading of mines actually took place during the years when the Carter Commission was considering the fiscal future of the industry. Had the concerns involved been small and obscure, the commission would have obtained little lethal ammunition. Instead, they were two of the largest and longest-established companies.

The first case had to do with a high-grade polymetallic mine in northern Quebec, namely Lake Dufault, a subsidiary of Falconbridge Nickel Mines. It started production in August 1964 and earned $0.57 per share during the next few months, $4.10 in 1965, $4.56 in 1966 and $3.57 in 1967. The next year it was evident that the honeymoon was over, the tax free period ended because net profits dropped to $0.68 per share. It is interesting that traditional investors in mining stocks, a rather sizeable percentage of the Canadian population in those days, were not fooled for one minute. During the three years of savage exploitation of the mine, the stock sold at three times earnings - this at a time when shares were usually considered to be undervalued if offered at a figure less than 15 times earnings. Also, the Canadian Mines Handbook (1966 and 1969/1970) gave the reserves in 1964 as 2,189,500 tons grading 3.9 percent copper, 7.0 percent zinc, 2 ounces per ton of silver and 0.03 ounce of gold. For 1968, the reserves were revealed to be only 909,000 tons of 1.6 percent copper, 1.56 percent zinc, 0.6 ounce of silver and 0.01 ounce of gold. Evidently, too much had been "creamed off." In 1971, Lake Dufault was officially amalgamated into Falconbridge, was no longer listed on the Toronto Stock Exchange, and thus disappeared from public view. Was this corporate streamlining or corporate embarassment?

More ammunition was given to the commission by the acrobatics which took place before and after the opening up of Pine Point Mines, an operation in the Northwest Territories owned by the world's largest lead-zinc-silver producer, Consolidated Mining and Smelting Corporation. The mine entered its tax-free period in November 1965. According to the Canadian Mines Handbook (1969/1970) by the time it expired, 1,350,000 tons of high-grade ore had been shipped to Kimberley and Trail in British Columbia and Kellogg, Idaho for "testing" purposes. I do not have specific and personal knowledge of the mineralogy of Pine Point, but it is almost certain to involve clean, very easily separated galena and sphalerite. However, let us assume for the sake of fairness that there were complexities. Testing could then involve pilot plant treatment of about 25 tons per day for at most a few months; or, under extraordinary circumstances, there might be semi-industrial runs involving a few thousand tons. Still there is no way that a throughput of well over a million tons grading about 50 percent lead-zinc combined can be viewed as anything but an outstanding example of "making hay while the sun shines." However, it should be pointed out that the geometry of the deposit was favorable for high-grading as the lead-zinc values decrease with depth and it is, an open-pit operation. Accordingly, from a technical viewpoint, one could seek exculpation by pointing out that you open pit from the top down. But a very strong case can be made that, had there not been a tax holiday, long-range economic considerations would have dictated extractive procedures to preserve rather than exploit maximum quantities of the richest material.

In short, the Carter Commission did not have any difficulty in putting an end to the basic support of new mining developments in Canada. The mining industry itself laid a red carpet in the direction of abolition of tax holidays.

With respect to percentage depletion, Thomas J. O'Neil points out that in the United States most of the rationale for granting depletion allowances does not stand up to critical analysis. But he does say that "The capital recovery argument is the only sound theoretical defense of mineral depletion allowance. The pragmatist, however, can argue that mineral industry financing is what it is: i.e., the structure of a whole industry is built around the depletion allowance and changing the rules of the game would have very damaging effects to the entire economy. It is ironic that at the very time when we are being threatened with mineral shortages of unprecedented dimensions, Congress is contemplating action which would intensify the problem. However valid the return-of-capital position might be the critic concerned with 'fairness of tax policy' will likely remain unsatisfied. Fortunately, the return-of-capital argument places the mineral depletion on a firm theoretical foundation. If one accepts depreciation as a legitimate tax reduction, one must acknowledge the legitimacy of the depletion deduction" (October, 1974).

This is all well and good, but the analogy is perhaps a bit unfortunate. Public policy does not permit depreciating a piece of equipment indefinitely, whereas with percentage depletion a large number of mines have recovered capital several times over. When this matter is mentioned, the industry invariably claims that it needs the extra revenue to explore for the new mines which will be required to ensure a life for the individual concerns far into the future. This is an argument which the Carter Commission quite adroitly pounced upon to justify a major change in the depletion concept; instead of an across-the-board percentage allowance seemingly ad infinitum, the depletion credit was to be tied directly to actual outlays for seeking and developing new mines.

FEDERAL LEGISLATION

By and large, the legislation which followed the report of the Carter Commission in 1972 was sound. The industry, of course, did not like it, but on the whole, its actions made changes mandatory. The unfortunate aspect is that Pandora's box was opened. For perhaps 40 years mining in Canada had operated under a basically unchanging set of rules which were clear-cut and simple. During this time the rate of discovery and bringing into production of new properties was unmatched anywhere in the world with the possible exception of the Soviet Union. But with the enactment into law of changes stemming largely from Carter, near chaos engulfed the mining industry. Since 1972 there has been more backing and filling: granting this concession or imposing that tribute today, annulling it tomorrow, and then restoring it a year later. This has not been exactly conducive to creating an aura of confidence. On the contrary, it has signaled to those who finance mining enterprises that danger lies ahead because those in power have not been able to agree upon or to impose a coherent philosophy.

The broad terms of the Federal Income Tax Act which came into force in January 1972 were as follows:

1. Termination in December 1973 of the three-year tax-exempt period with new mines.
2. Automatic percentage depletion to be replaced by earned depletion with a rate of thirty-three and one-third percent, effective January 1977 (later changed to May1974 at the reduced rate of 25 percent).
3. Accelerated capital cost allowance for new mines and major expansions. This allowance, taken together with the right to deduct exploration and development costs as incurred, meant that a new mine would pay no federal income tax until capital investment had been recovered. In 1974 this was changed with respect to development expenses which became deductible at the rate of only 30 percent of the unclaimed balance.
4. Tax rate of 50 percent on production profits, falling one point annually to 46 percent in 1976. Abatement on account of taxes and royalties paid to the provinces to amount to 25 percentage points after 1976, resulting in an effective federal tax rate of 21 percent should the provinces take full advantages.
5. Provincial mining taxes and royalties not deductible after 1976 (later changed to May 1974) and these taxes and royalties cannot be treated as operating costs.
6. Prospectors and their financial backers no longer exempted from income tax on receipts from sale of a mining property or capital gains taxes on any shares which may have been part of the payment.

There is no need to comment on the articles dealing with depletion allowances and tax-exempt periods. As for accelerated capital cost allowance, this provision goes a long way toward stilling the doubts of potential lenders and removes all incentive to high-grade. Article 4, on the other hand, is loaded with dynamite with short fuses. Before the Federal Income Tax Act of 1972, abatement of provincial taxation amounted to ten percent. Nevertheless, seven out of the ten provinces had set higher corporate tax rates. The allowing the abatement of 15 additional percentage points in the act was an open invitation fanning the greed of the provinces. As I shall illustrate later, several of the provinces evinced an inordinate degree of enthusiasm.

As for taxes, there was to have been an effective federal tax rate of 21 percent after 1976. In November 1974 the overall rate was fixed at 50 percent, replacing the gradual reduction annually to 46 percent. The resource abatement which was to have come in 1976, instead was made effective retroactively to May 1974, resulting in the effective federal tax on mining companies amounting to 25 percent of net profits. The next budget in June 1975 dropped the rate back to 46 percent effective January 1976. The extra 15 percentage points of abatement were withdrawn and thus the federal tax rate became 36 percent. A new deduction, the resource allowance, was introduced. It amounts to 25 percent of resource production profits. All of these changes resulted in increasing the federal take to 27 percent. However, this rate is effectively lower if the company has interest charges. exploration and development expenditures as these are deducted after the resource allowance.

The prospector and promoter have been vital elements in the growth of the Canadian mining industry. It is questionable that many are left. However, this is understandable. With the new regulations, 50 percent of capital gains were treated as income and taxed at the applicable rate. Later, the high rates of tax were somewhat ameliorated by an amendment to the Federal Income Tax Act in 1973 to permit investment of the proceeds from the sale of mining properties in income-averaging annuities. Nevertheless, the main drive to mine-finding in other days - the chance to strike it rich - has been greatly weakened. The game will simply not be worthwhile if a government which has shared none of the hardship and high risk will obtain a large part of the fruits when very long shot bets pay off handsomely.

If the foregoing leaves the reader somewhat confused, this is in part a function of trying to write a chapter which contains fewer words than the Encyclopaedia Britannica. More important, bemusement is generated by the evident lack of understanding of what the mining industry is all about on the part of those who vote a new set of legislation almost every year. Apparently the government does not realize that one of the principal inhibiters of investment is uncertainty. Government exactions can be tough but they can usually be lived with and adjustments made if they all only changed gradually and infrequently, which is the reverse of what has happened in Canada during the last few years. Sound corporate planning cannot be done on an annual basis. Accordingly, if every new budget which comes down is likely to be full of unpleasant surprises, the prudent mining house will develop a marked disinclination to undertake any new investment beyond that required to stay in operation.

PROVINCIAL LEGISLATION

Ontario

Effective April 1974, the Ontario Mining Tax was replaced by a graduated rate schedule which treats associated companies as one. The profits of a subsidiary of International Nickel in Guatemala are dropped into the pot but any mining taxes or royalties paid to Guatemala are not allowed as deductions in computing taxable income for Ontario purposes.

In announcing the legislation, Ontario cited such things as "sharply higher metal prices and windfall gains for the mining industry in Ontario" and the need to secure for the people of Ontario a "higher return from our natural resources" (Brown, July 1974).

A couple of irreverent questions come to mind. Why is it that the most basic of all resources, land, has been exploited ruthlessly from the beginning of pastoral man without intervention by governments except in rare instances? Then there is the matter of that very evil thing called the windfall profit. Of course, modern economic and social theory dictates that it should not be countenanced. But why is it that the excess profits which are skimmed off are not put in escrow for that year or years, in the near term, when the mine has great difficulty in keeping its head above water. The answers are very simple. No politician will risk a confrontation with farmers - there are too many of them. And no politician is unaware that jousting with big business (which mining must be to survive) will enhance his image as a "defender of the heritage of the people." It is of course quite irrelevant that the "people" and their elected "defenders" are almost totally ignorant of the characteristics of that "heritage" and what has to be done to make the most of it.

As for the new tax regime, Alfred Powis, president of Noranda Mines, has said: "The savage new Ontario graduated taxes mean that Noranda would pay a rate of over 60 percent on profits from new mines development in the province, compared with around 40 percent in the case of manufacturing and processing industries. At the root of the problem is the repeated assertion which cannot be supported by sound analysis that the industry earns excessive profits from the extraction of non-renewable resources belonging to the people and that an inadequate proportion of these profits are returned to the people in the form of taxes" (September 1974).

M.C. Bartley, (September 30, 1976) president of the Canadian Mining and Metallurgical Institute, pointed out that no new iron ore mines have been developed in Canada since 1968 despite vast resources. Yet, three Canadian steel companies have taken equity positions in Minnesota taconites which will result in imports into Canada of five million tons of pellets, equal to 40 percent of present consumption. The decision to go abroad was made strictly on the basis of the cost of the ore and return on investment. For Canadian ore to be competitive with the taconites of Michigan and Minnesota would require a reduction in the CIF price of about 20 percent, or $4.00. While an iron rich country is turning to inferior foreign sources, four of Ontario's iron ore mines will close down during the next five years. Of course, more than taxes are involved here. The cost of labor in relation to productivity is a factor of increasing concern. Nevertheless, the heavy impositions of governments which reduce return on investment to a marginal point are, to put it mildly, not conducive to growth.

Saskatchewan

Table 9.1 depicts the horror story of the taxation of potash output in Saskatchewan.

TABLE 9.1. Taxation of the Potash-Producing Industry in Saskatchewan

Year	Value of Output (a) (in thousands of Canadian dollars)	Provincial Taxation (b) (in thousands of Canadian dollars)	Ratio of Taxation (c) percent
1970	111,884	1,976	1.8
1971	145,966	2,554	1.7
1972	135,513	5,253	3.9
1973	194,621	9,904	5.1
1974	312,544	53,343	17.1
1975	355,000	105,783	29.8

(a) Calendar year

(b) Fiscal year beginning with year shown.

(c) Taxation to gross value of output.

Source: Annual Reports of the Department of Mineral Resources, Province of Saskatchewan, for the fiscal years ending March 31, 1971 to March 31, 1976.

How do you justify a tax of 29.8 percent on gross? The industry as a whole does not net half that percentage. The answer is more than obvious. You impose such taxation when you intend to confiscate an industry or, more precisely, to make it more than amenable to a takeover. Malcolm Scott, general manager of Central Canada Potash, claims that "the government has been systematically reducing producers' equity by onerous taxation policies to the point where nobody but the government would be interested in taking on the load - they are the only buyers in the market" (Engineering and Mining Journal, December 1975.) One company has already thrown in the sponge. In August 1976 a definitive agreement was executed for the sale of the mine and mill of the Duval Corporation of Canada to a public entity, the Potash Corporation of Saskatchewan, for $128.5 million, or about $0.30 per ton of proved reserves.

Panorama, the monthly in-house journal of Noranda Mines Limited reports on the tax situation with its subsidiary Central Canada Potash. "In the fiscal year ending June 30, total tax payments of $26,777,000 turned an operating profit of $19,603,000 into a net loss of $7,174,000. The situation is not new. The total of all taxes levied and paid to governments from the beginning of production in July 1970 to the end of the fiscal year ending June 30, 1978 has amounted to $80,366,000 of which $60,633,000 was paid to the Province of Saskatchewan.

In addition, a further $9,750,000 in taxes, fees, assessments and penalties claimed by Saskatchewan is disputed by the Company and has not yet been paid. The total tax burden exceeds $20.00 per ton of muriate produced, representing almost half the selling price." (Panorama) 1978).

Listed below are the successive revenue-producing acts which, in the final analysis, have been clearly aimed at nationalization. (One is tempted to say "provincialization" but refrains in sympathy for a noble language which is being bastardized at too rapid a rate.)

1. Subsurface Minerals Regulation (1960) - Royalty rate on potash varying from 4.25 percent to 9 percent depending on grade of ore and assessed on producer's net sales price.

2. Mineral Disposition Regulations (1961) - Based on income from mining operations, e.g., nothing for those with operations of up to $25,000 and 9 to 12.5 percent for those with over $500,000. In the latter category, the 9 percent rate is imposed on new mines and the 12.5 percent on existing ones.

3. Potash Proration Fee (1972) - $1.20 per ton of muriate produced.

4. Potash Reserve Tax (1974) - Assessed on gross revenues, the tax is not imposed until the price of potash rises above $35.50 a ton. At or above $90.00, it amounts to about 38 percent. It is calculated by a complicated formula which, by definition, implies bureaucratic discretion and attendant inequity involving, in addition to selling price, the productive capacity of the mine, the grade of ore and the capital cost. It is believed that this tax represents only the second attempt to tax reserves in the history of Canadian mining - the first on iron in British Columbia, was soon repealed - although it will probably not be the last.

A new royalty scheduled for uranium was announced early in 1976. It involves three percent on gross (which is equivalent to about 20 percent on net) plus a graduated royalty scaled as follows:

Ratio of Operating Profit/Capital Investment (percent)	Tax Rate (percent)
Less than 15	0
15 - 25	15
25 - 45	30
More than 45	50

Virtually nothing would be left over in prosperous years which all successful mining enterprises must enjoy if only to bridge over the bad times which inevitably follow. What makes this matter highly ironical is the fact that the only producing uranium mine in Saskatchewan belongs to the Crown Corporation, Eldorado Nuclear which is totally owned by the federal government. The tail wagging the dog? No, it is far worse! The province is boldly saying to the federal government: "We will take all of the revenues from mining, including those of your own enterprises." The fact that, in the process, an industry may be destroyed has undoubtedly occurred to and has been expeditiously ignored by the politicians issuing decrees. It is harsh to say, but Saskatchewan provides an outstanding example of a political philosophy in this country which seeks not positive results but instead sensational causes.

Early in the Seventies, Saskatchewan imposed a prorationing scheme with potash involving a floor price of $33.50 per ton and production quotas. Naturally, this meant curtailment of production, and producers in the rest of the world profited very nicely from that province's inept manipulations. Then, in 1975, the courts found that prorationing was unconstitutional in that it amounted to an attempt to dictate the terms of trade internationally, a function to be exercised only by the federal government.

This setback was not taken lightly. The government announced its intent to obtain by compulsory purchase the "commanding share" of the industry. The rationale for nationalization was that the mining companies refused to expand production. A spokesman for the industry portrayed the other side of the coin: "Virtually every producer in the province has gone to the government with proposals for expansion if they could be given some form of tax relief and all were turned down. Now Blakeney* says unless Saskatchewan increases production, other countries, like Russia, will fill the void. We have been telling him that for years" (Engineering and Mining Journal, December, 1975). It is interesting to note in this connection that there were expansion plans on

* The Premier of Saskatchewan

the order of $200 million for 1975 which were shelved pending court action by the Potash Producers Association to have the reserve tax declared unconstitutional.

British Columbia

British Columbia, under the rule of the National Democratic Party (NDP), clearly established a "blood brother" relationship with Saskatchewan. During its mercifully short tenure which ended in 1875 the NDP inflicted upon the mining industry taxation legislation which was not only severely punitive, but also indicative of either abysmal ignorance or, more probably, the intent to nationalize. To earth scientists concerned about the health of the mining sector, it would be enormously reassuring to know that the people of British Columbia rejected NDP because of its shameful treatment of the industry. Unfortunately, things are not as logical in this complex and confused world. It has been postulated that the fastest gun involved in shooting down NDP was the teachers' association. There was confrontation which found the government and the educators glaring at each other over the barricades. The latter do not have the prestige, power or economic muscles of mining executives; but they do have numbers and recognized merit, and they are listened to in the inner communities. The mine manager speaks usually only to God, but regrettably, legislators do not consult with the Supreme Diety when writing laws.

The first mining legislation enacted was the Mining Tax Act which exacts 17.5 percent (effective January 1, 1976) on net income in excess of $10,000 derived from mining operations. Then came the Mineral Land Tax Act which imposes an acreage tax of no great severity. But the pièce de résistance was the Mineral Royalties Act. This was passed in June 1974 and applied initially to only four minerals: copper, gold, silver and molybdenum. It involved a basic royalty of 2.5 percent in 1974, thereafter 5 percent on gross revenues. As pointed out earlier, such seemingly petty figures can amount to as much as one-third of net profit. Far more formidable, a super-royalty of 50 percent was imposed on return above 120 percent of a basic value for the mineral concerned - a value to be determined by the cabinet. For copper the basic value was set at 58 cents per pound. Thus, anytime the price of copper rose above 70 cents, the government grabbed 50 percent of the difference. With gold, when the price rose above $99.00 per ounce the excess was to be similarly shared fifty-fifty. Later (early in 1975), eight other minerals were added to the list - asbestos, cadmium, cobalt, gypsum, lead, mercury, nickel and zinc - and the confusion of the initial legislation was compounded. There was simply no apparent rhyme or reason. Almost all the lead and zinc comes from one mine in British Columbia - the Sullivan of Consolidated Mining and Smelting Corporation. Yet, relating basic values multiplied by 120 percent and the minerals prices quoted at that time by the Engineering and Mining Journal (February 1975) produces very bemusing results. With lead, 120 percent of basic value was $0.156 per pound, whereas the Engineering and Mining Journal (February 1975) quote was $0.245. For zinc, which comes from the same piece of rock, the corresponding figures were $0.14 and $0.39.

Two points stand out. First, basic values were set so low in general as to ensure "windfall taxes" for the government - prohibitive taxes, actually. Second, it is virtually impossible to deduce the reasoning, if there was any, behind the establishment of rates. One would assume that a logical system of taxation goes straight across the board - such and such percentage for all. On the contrary, the authorities in British Columbia managed to produce a hodge-podge of rates that was bewildering. In general, the slant was in favor of the government but never with any consistency, never in any way reflecting well-reasoned policy.

The Mineral Royalties Act should rank high among the most irresponsible legislation ever enacted in Canada or be viewed as part of a sinister design to expropriate an industry at a minimal cost. As R.D. Brown noted, "The new mineral royalties legislation provides a basis of taxation which is unrelated to mining profits; instead it is largely dependent on the international price for the particular mineral. The marginal rate of federal and provincial taxes and royalties on additional income accruing to the producer as a result of increases in the international price can be as high as 98 percent" (March 1976). In a footnote, he added that marginal rates may be even more than 100 percent if cost increases are not matched with increases in the basic mineral values used in calculating the super-royalty.

The act in question had a quite dramatic impact in many spheres. To take an example from exploration, according to the Canadian Diamond Drilling Association, during the first quarter of 1974, 256,514 feet of exploratory drilling took place. In the corresponding quarter of the next year, the total was 2,400 feet - an exceedingly dramatic cutback of more than 99 percent. Similarly indicative is the statement in the Engineering and Mining Journal, that "From 1972 through 1974, development spending dropped by 98 percent while claim staking fell 80 percent - to the lowest point in more than a decade" (July 1975).

The government which replaced the National Democratic Party offered new hope to a beleaguered mining industry. In the first instance it repealed the Mineral Royalties Act. All well and good. But then came the words of the new Minister of Mines and Petroleum Resources: "Our province has put its taxation on the right basis - a profit based tax. From a low of 48 percent for new mines with capital expansion and write offs to a high of 56 percent for old non-expanding mines, this is in line with other manufacturing and resource industries. A fair sharing basis still has to be worked out with the federal government but there is no longer any unfair discrimination against mining companies in our province" (Chabot, January 11, 1977). One wonders at the logic of the proposition that tax rates of 48 to 56 percent - even though they do include the federal impost - constitute an end to unfair discrimination. Admittedly, when compared to the brutality of the Mineral Royalties Act, the change for the better was striking. Nevertheless, once again the high-risk nature of finding, operating and making a competitive profit from mines was conveniently ignored.

CONCLUSION

The mining industry has been badly abused in recent years. That in some respects it has deserved such treatment is irrelevant. What counts is that Canada, more than most nations, is critically dependent upon a healthy mining sector. It is therefore unfortunate that by and large the industry's public relations record can at best be described as inept. The public knows virtually nothing about the distinctive characteristics of mining if only because the industry has never deigned to establish a dialogue. The main lobby, the Canadian Metal Mining Association, gets press coverage usually when it is parading to Ottawa to protest this foreign tariff or that control, and it is usually portrayed as a petitioner for special interests rather than as a constructive force. In addition, mining companies rarely publicize their failures. No mention is made of the five hundred or more raw prospects upon which work - frequently quite substantial work - has to be done in order to find the one property which will become productive. Except in the technical journals, it is not revealed that 30 to 60 million dollars must be spent to find and outline a major ore body today. The general public is fed only the good news, the success which makes top management appear brilliant and omnipotent, and therefore it cannot be blamed if it sees a very distorted picture. The politicians may occasionally know better but the basic principle of holding onto one's political seat being to appeal to the lowest common denominator - the majority of the voters, their actions will tend to mirror the ignorance, prejudices and envy of the electorate.

Too often the industry has shown little vision and statesmanship. The debacle in British Columbia was made easy for the politicians because the copper and/or molybdenum producers were very vulnerable. In 1973, copper production in that province was to 350,000 tons and molybdenum production was 30,000 tons. It is a national disgrace that every pound was exported in concentrates and no further working was done. It is unconscionable that copper production of vast magnitude in British Columbia was not converted to refined form which would have improved Canada's balance of trade in 1973 by about $60 million. Of course, it is evident that the tune was called abroad, chiefly by Japan. But during most of the Sixties there was a seller's market. Nevertheless, the individual mining concerns capitulated without much of a struggle either to obtain a quick loan from abroad or to be able to enter the banker's chambers with a forward sales contract firmly in hand. That such contracts sometimes turn out to be mere scraps of paper when prices plummet apparently occurred to few, if any, participants in what was one of the largest mining developments in history and one of the most heedless sellouts of natural resources imaginable. Certainly, these days, no least developed country in the world would permit exports of several hundred thousand tons of copper annually in any but blister or refined form.

Another example can be provided by the Pine Point operation referred to earlier. Its concentrates of lead and zinc are very high grade, uncontaminated with impurities which complicate smelting, and they belong to the world's largest producer of these metals. But, because the very purity of the concentrates is not easily compatible with other concentrates flowing into the Trail Smelter, Pine Point production traditionally has been smelted in Japan. When one considers that annual output amounts to several hundred thousand tons of lead and zinc combined, it is not difficult to realize that the government probably felt that the company, by paying scant attention to the interests of the province and the nation, was acting in a rather cavalier manner. But I feel very strongly that the actions taken by the government were not aimed so much at chastizing errant free enterprise as at back-door nationalization by impoverishment.

The federal government is apparently unable to resolve the conflict with the provinces so as to establish a rational tax regime. And it appears to be ready to assume, with only the slightest provocation, a nationalistic, not to say chauvinistic, posture - an attitude which does not imbue foreign capital with great desire to participate in building the future of Canadian mining - or other industries for that matter. On the other hand, the Department of Energy, Mines and Resources recently found that about $250 million (in constant 1974 dollars) will be needed annually for exploration over the next 25 years simply to maintain Canada's position as a producer and exporter. It will be no mean accomplishment to lay hands on these funds because they amount to about three times what has been spent per annum in seeking mines during the past 20 years. Far more impressive is the estimate of $900 million annually to expand existing operations and develop new metal mines. This is probably at least two to three times what is currently being invested in new productive facilities. Not included in these figures are massive outlays for infrastructure, transportation, downstream processing, etc. Where will the money come from? Foreign capital is more than wary of Canadian mineral investments; it is downright terrified. The small investor who traditionally was a potent force collectively goes to the race track or buys lottery tickets these days. Institutional funds are probably being placed everywhere except in mining issues.

Under these circumstances - and with remembrance of apparent attempts to nationalize the mines of British Columbia by taxation and the on-going program in Saskatchewan, one can envision a steady progression in the direction of the national and/or provincial government becoming the proprietor of Canadian mining. Let us pay some heed to the only slightly veiled menaces of Leo Bernier, a former Minister of Natural Resources for the Province of Ontario, who noted that the mineral industry in Canada "is undergoing the greatest scrutiny that it has ever received from governments and the public, and in some quarters it is under attack. Seldom a day goes by that one of your critics does not suggest the immediate advisability of nationalizing all metal mines in Canada or placing such a tax burden on you that it would be difficult if not impossible for you to survive in your present form" (Engineering and Mining Journal, May 1973). If there is nationalization, it will threaten to infect the entire country. In the end, the federal government, if powerful enough, would probably have to step in to renationalize so as to bring the entire industry under its wings. While admitting that such intervention could become necessary, I fervently hope that we will not see it.

On a very limited basis, there has been experience for more than a quarter of a century with government ownership of mines in Canada. It has not been unfavorable. But, to what extent is it representative? The experience involved Eldorado Uranium Limited, a company initially formed to provide leadership in the exploration, production, purchase and marketing of nuclear material early in World War II. It has operated important mines for 30-odd years, and operated well, for good reasons. In the first place, it offered salaries at least competitive with private enterprise. In addition, its board of directors was composed largely of men from the industry rather than civil servants. And both presidents and directors were outstanding. I doubt that the experience of Eldorado Nuclear can be repeated. First, whereas Ottawa could afford to give virtually autonomous status to a medium-sized company such as Eldorado, it would not grant such independence to an entire industry. Second, most people from the private sector would stay on from necessity rather than choice and in the process be lulled to sleep by pension plans and job tenure. Under these circumstances there is rarely any challenge, and, without that, there is at best mediocrity. The next item to consider is the matter of nepotism and appointments totally dictated by politics, definitely not factors in profitable equations. Then there is the matter of disposal of earnings. Sound practice dictates that a large part of these be reinvested in the business in order to keep it strong and growing. Every impulse of government is to syphon off as much as possible to bolster up whatever program is believed to be desired by those who are going to put their mark on a ballot shortly.

It is too early to document statistically the havoc created by recent acts of government. The time span is too short and the times are too turbulent. To assess the relative force of the blows which have buffeted the industry during recent years will require more perspective than is obtainable today. But I propose that unless there are decided changes in the Canadian political climate, the story which will emerge by the early Eighties may turn out to be a tale of horror better not read during the witching hours.

In my opinion, what has been happening in recent years in Saskatchewan and British Columbia and which is mirrored to a lesser degree in other provinces poses a greater threat to the survival of Canada as a nation than the separatist movement in Quebec. Clear-cut plans of expropriating by taxation of a very basic industry have been evident. They have brought the copper industry in British Columbia to its knees and it is by no means certain that the industry will be able to straighten up. The potash producers in Saskatchewan are in the process of being taken over at bargain-basement prices. A series of fundamental questions arise: Can a country hold together without a controlling ideology? Can provinces, petty little principalities though they may be, impose doctrine in contravention of the central power? Can a society more or less wedded to free enterprise contain, within it, sovereign entities dedicated to its destruction?

POSTSCRIPT

In February 1978, considerable evidence emerged which reveals that the remarks made at the end of this chapter amount to understatement of classic proportions. The Mining Journal chronicles a bitter battle in Saskatchewan concerning prorationing of potash production:

> Consistently, both before a Saskatchewan Queen's Bench judge and, subsequently, before the Saskatchewan Court of Appeal, CCP (Central Canada Potash Limited) has argued that the Saskatchewan Government had no right effectively to prevent the export of CCP potash to the U.S. under a long term contract. That contract with CF Industries, a partner with Noranda in CCP - was a major raison d'etre in the Colonsay mine.
>
> CCP argued that it was unfair that, having secured a long term market for its output before coming into production, it should be penalized because other companies had not, and that, since regulation of external trade is a federal rather than a provincial government function, the prohibition of its exports was unconstitutional. This latter thesis was supported by the federal government.
>
> Whereas, the Queen's Bench judge found in favour of CCP and awarded the company $C1.5 in damages, the Saskatchewan Court of Appeal decided that the provincial government was within its rights to control the production of a provincial resource and withdrew the award (Mining Journal, December 30, 1977).

The article also points out that the case has been argued recently before the Canadian Supreme Court which so far has reserved judgment. However, it would be a mistake to assume that a ruling in favor of CCP by the court would put an end once and for all to control of production for export by the provinces. The Federal government feels that regulation of external trade is very much within its jurisdiction. But whether it will be able to assert its authority is open to question. The Supreme Court of Canada does not bear much resemblance in power terms to the one which sits in Washington. This assertion is well illustrated by recent tidings conveyed by the Engineering and Mining Journal:

> A decision by the Canadian Supreme Court late in November declaring Saskatchewan's income tax and royalty surcharge on oil production unconstitutional has given the province's potash producers new confidence that their legal challenge of the provincial potash reserve tax and prorationing fees will prove successful. At stake is more than $120 million in tax levies. Counsel for the 11 potash producers seeking to overturn the reserve tax described the Court decision as "precedent-setting from a legal point of view...the latest word from the Supreme Court about the imposition of indirect taxation and what constitutes direct and indirect taxation."
>
> However, in response, the Saskatchewan Government quickly introduced a new oil income tax proposal aimed at countering the Supreme Court decision. The new tax would be retroactive to January 1, 1974, but the tax rates would be set by future legislation. Both opposition parties in Saskatchewan, the Liberals and the Progressive Conservatives, supported the governing socialist New Democratic Party in opposing rebate of tax payments. (January, 1978).

This is, in effect, open defiance of the federal government in Ottawa! While one sees civil war in Canada as being inconceivable in the late twentieth century, close analogy can be drawn to events in the United States a century and more ago. As the Constitution was being hammered out in 1787, by and large the people thought in terms of their state rather than the nation. This belief remained strong in the South and in the end led to war. Fortunately, the framers of the Constitution ensured a strong union rather than Balkanization by setting up three independent but interdependent bodies at the national level - the Congress, the Presidency and the Supreme Court - and ensured centralization of power. Canada on the other hand has the Federal Parliament, a Senate which is a club as impotent as the House of Lords to which the party faithful are retired, and a Supreme Court with a quite limited mandate. Pitted against this trio are provinces which appear ever more in adversary roles and too frequently with impunity. Around the village square, it is, no doubt, good politics. The only trouble is that the country could easily blow apart.

10 The Prevention of Ghost Mining Towns in Arid or Desert Zones*

We all have a clear idea of what ghost towns look like, chiefly because they have been prominently featured in Western movies: gaunt, weatherbeaten, crumbling buildings, tumbleweed rolling down a dirt road, the spirits of people long departed still faintly haunting the place ... There is a sadness to the scene stemming from realization that it represents loss, waste, disruption of lives and destruction of a community. Once thriving businesses almost overnight become valueless - the swinging doors of Black Mac's Drinking and Dancing Emporium swing only with the wind. The dentist, the doctor, the lawyer - all have to relocate and start once again the arduous struggle to build up a practice. As businesses close down and services are cut, those who might care to stay on have no choice but to join the trek. All lose to a certain degree; some lose almost totally.

Very often, ghost towns have been a by-product of the exhaustion of mines. It is these ghost towns I shall talk about in this chapter. Specifically, I shall address the ways of ameliorating or eliminating the waste and sociological stress involved in the death of such towns.

POSSIBLE APPROACHES

Trailer Camps

At first glance, the establishment of trailer parks appears to offer interesting possibilities. Certainly, it would save money and there

---

* Adapted from paper given at the Conference on Alternative Strategies for Desert Development and Management, Sacramento, California, May 28 - June 10, 1977. The copyright is reserved by the United Nations Institute for Training and Research.

would be salvage value when the mine closed down. However, for a number of reasons, I believe that meeting housing needs with trailers would be the worst possible way to try to solve the problem. First of all, savings would not be as substantial as one might think. Even a trailer park has to be supplied with water, electricity, sewage disposal and roads. In addition, in a day when it is difficult in so many places to attract and hold good miners, a large part of the cost of establishing a new community is in the provision of common services and facilities. Shopping center, bank, specialty stores, hospital, schools, playing fields, parks, arena and community center - these are costly, particularly as they have to be of high quality.

Second, far too many people who would arrive in a mobile home would be drifters, staying a few months to get a stake and then going on again.

Third, by and large trailer living involves more than a little psychological stress, not to mention plain discomfort from over-crowding. In general, an unhappy family life does not produce a good worker.

Last but not least, for many people, limitations of space would lead to the decision to limit the number of children in a family. Thus, the powerful anchor of children in school, playing Little League baseball or soccer, would be lost.

Planned Development of a Mining Area

Where a number of ore bodies are located within a relatively small area, a centralized community could, in theory, be created to service each in turn: as Mine A becomes exhausted, Mine B swings into action, then Mine C, and so on. Thus, the life of the area would be greatly prolonged. It could be done, and it probably is being done, in the Soviet Union among other places, under the control exerted by the centrally planned economies. But where there are still vestiges of laissez faire, one questions whether the owners of Mines B and C will be willing to defer commencement of profits by 20 or 40 years, as the case may be. Also, in the developing world, even with mines which are wholly owned by governments, the pressures to maximize near-term gains are very real and strong; the problems confronting governments are so immense and complex that tomorrow will almost always be sacrificed before the stark exigencies of today.

Reducing the Scale of Operations

The question may be posed as to whether it would be advantageous from a national as well as from a company point of view to restrict the rate of mining to a moderate scale so as to extend the life of the mine to, say, 60 to 100 years instead of perhaps 15 to 25 years. In reality, most mining companies pursue a policy of maximizing production if only because of uncertainty as to what the future holds in a very risky world.

In other words, get it while the getting is good. Increasingly, though, the company has no choice in this respect for the following reasons:

1. In a growing number of places, governments are now offering exploitation concessions of as little as 15 years, rarely more than 25. That the agreement offers possibilities of extensions has to be assigned a book value of about $1.00 by prudent management.
2. Governments have numerous other ways of applying pressure for speeding up rather than stretching out operations.
3. The characteristics of the ore body, particularly grade, are becoming more and more the major determinants of scale of operations.
4. Banks and other sources of financing will usually view with a decidedly jaundiced eye any scheme which increases costs per unit of output.
5. In the United States, a number of states tax in-ground reserves, in part to obtain revenue, but also in order to prevent the mining companies from sitting on reserves.

From the viewpoint of governments, extending the life of mines is usually attractive only when the social costs involved in closure outweigh the loss of revenues. Extreme examples of such instances are:

a. Continuing operations of the Portobello gold mine in Ecuador about 20 years after the foreign owners walked away because reserves were almost exhausted and it simply was not an economic proposition anymore. Today, operations still go on, heavily subsidized by government, of course, because a community numbering several thousands is involved. Annual losses in recent years have amounted to a million dollars or more, which is no mean price to pay for sidestepping the creation of a ghost town in a poor country.
b. Howard-Goldsmith described the ultimate in closed circuits, probably still operating at Colomb-Beshar which lies deep in the Algerian Sahara. In the late 1950s, the French developed a coal mine to fuel an on-site generating stations which was important to a nation in desperate need of energy. Before the transmission line was built, great reserves of oil and gas were discovered. Colomb-Beshar no longer had any economic justification, but as late as 1970, the mine was being worked simply to supply the town with power and provide employment (1971).

The main advantages of extending the life of mines for the mining companies would be:

a. Reducing scale would often mean cutting financial requirements to the point where funding could be provided from internal sources. This, however, would be attractive solely in times of tight money, when financing can be obtained only at exorbitant interest rates. Normally, the mining concerns seek the maximum percentage of loan capital so as to increase the leverage of the equity holders.

b. Mining personnel would be offered the opportunity of employment throughout their working life and thus labor turnover and training of new personnel would be considerably reduced. However, as will be shown later, there are other means of ensuring these desirable results.

Use of Already Existing Towns

Where the mine is located in a physically hostile setting and yet lies within reasonable distance - let us say, 50 miles or less from an established town - serious consideration should be given to constructing only those facilities required for mining operations on the site. The workforce would be domiciled and serviced in the nearby community. Generally, people would arrive at their place of work by vehicular transportation. In other cases, where distances are well in excess of 50 miles, recourse can be had to a practice which is becoming common in northern Canada, namely, airlifting. Under this system, personnel work extremely hard seven days per week, twelve hours per day, with a commensurate rest period when returned to their permanent domicile.

Doing It the Cheap Way

"Historically, the company town has exerted maximum political, economic and social control over its inhabitants, while commonly offering minimal housing, facilities and amenities" (Coal Age, May 1975). In an earlier and usually brutally exploitive day, such towns were almost always an abomination even in the developed countries. They featured mean shacks, outdoor plumbing, sordid bunkhouses referred to generally as "ram pastures," and the company store, which was frequently used as an effective and profitable means to impose de facto enslavement. Where operations were so isolated that management had to live in proximity to the "serfs," contrasts were appalling. The psychological wounds visited upon both parties to such an unequal contract had to be extremely damaging. Nevertheless, these were the norms in many mining camps in more than a few parts of the industrial world up to World War II.

Conditions far worse prevail, in the Third World, particularly with mines which are publicly owned and operated. In places where families actually live in the mouths of abandoned adits or in adobe shacks constructed by the bare hands of the inhabitants, ghost towns have to represent the most minimal of financial losses.

Today, in those important mining countries where unions are strong and unemployment is low the converse is the case. A high standard of living has to be guaranteed in order to avoid a chronic shortage of skilled labor, not to mention administrative and supervisory personnel. The old bunkhouse has been replaced by comfortable apartments, well-planned and well-built separate houses, each on ample land, are emphasized. The company store, now a supermarket, frequently is a loss leader on the company's books. A former manager of a fairly large mine in northern Canada was asked whether mining operations, milling operations on labor relations were his major problem. None of these, he replied, but instead, the need to explain and defend the price increase of two cents on a quart of milk or a loaf of bread.

Doing it the cheap way is not a workable solution in the industrialized world. And one could hope with considerable feeling that with the new mines to come into production in the developing nations, there will be sufficient enlightenment on the part of governments and owners to ensure that the horrors of the past are not visited upon the future mining community. Whether the solution I shall propose later for eliminating ghost towns in the developed countries will work in the Third World is unfortunately questionable.

A PLANNED AND COORDINATED APPROACH

A number of proposals of doubtful merit have been made to date for reducing the social and financial losses represented by ghost towns. Instead, with imagination, vision and early planning, such towns can be relatively easily eliminated in almost all desert environments of the industrialized countries. No new mine should be permitted to come into production without taking measures to ensure a life of some sort for the community long beyond the day when the mines closes. After all, it is rather paradoxical to think of abandoned mining communities in the same deserts which are being inundated by immigrants seeking a place in the sun. As is evidenced by Arizona and other states with arid climates, the desert (excluding tundra, of course), is a very pleasant place in which to live. Sunshine, cool nights, relatively clean air (unless there is a smelter nearby), space and vistas - these things are golden to many who lived for long in "concrete jungles." Particularly to the old, a retirement home near Tucson, Arizona is infinitely preferable to braving the streets in which furtive figures slink as darkness descends, and enduring the grey dankness and cold of winter and the frequently stupifying summer heat of such places as New York and Washington.

What I am proposing is that new mining towns should be set up with the idea that in the end they will become at least partially retirement centers. To do so will require in the first instance good land. Fertile land will rarely be a problem because soils will not have been leached, but there will have to be space - space for the "good life" and for growth. Accordingly, in those instances where the mine is located in surroundings which are very rugged, consideration should be given to establishing the town in the nearest available place with gentle topography and to limiting construction at the mine to those edifices directly involved in the extraction and treatment processes. Of course, there has to be a limit in regard to distance, but there are, in any case, few ore bodies in the deserts which lie more than 40 or 50 miles from tracts suitable for community development.

The communities envisioned can be depicted by two recent examples of mining town planning. Neither are located in deserts but, as is the general rule with mines, they are not in the midst of a Garden of Eden.

At Colstrip, a major strip mining operation for coal has been set up in the range country of southern Montana. The housing breakdown is 226 single family detached houses, 60 multifamily units and 180 spaces in a mobile home park. (With regard to the latter, for reasons expressed earlier I hope that few people venture in.) Accommodations are of a high standard and rates are relatively low. Rent for a $30,000 house amounts to $210 per month including maintenance, water and sewer, garbage collection, landscaping, range and refrigerator and insurance upon the structure.

The philosophy behind not sparing cost in setting up the company town has been that it is recognized that poor living conditions lead to labor unrest whereas good housing and other facilities reduce turnover and improve morale which reflects favorably on job performance.

A considerably more interesting and advanced approach to creating something resembling the ideal company town, the town of the future, the town which can easily convert to a retirement place in the desert is the town of Leaf Rapids, a development of the Ruttan Mine in northern Manitoba (Clarke and Grimes, April, 1975).

The mine itself has an indicated life of 20 years. In that geological environment it is probable that it will survive in good health for considerably more time. The town site will be located 15 miles away. This is significant: typically, in the past everything was clustered around the headframe and men walked to work. In this instance, it is evident that aesthetics have pointed out that it is much more pleasant to live near the banks of a river and it is far easier to construct on bottom land than in the rocky bush.

From the beginning, the company realized that it did not have the in-house capacity to plan and construct a modern community sufficiently attractive and workable to lure people into the northern wilderness and keep them. Accordingly, the job was contracted to a major subdivision developer. Initially, 600 single-family units on lots measuring at least 60 by 100 feet and 300 multifamily units were planned. All electrical wiring was to be buried.

In 1971 a startling event took place. The Manitoba Government decided to build the town itself, using the town developer referred to above. This was not an indication that the government eventually wanted to be the proud owner of a ghost town but, instead, recognition of a trend which is bound to grow stronger, namely, homes owned largely by employees and facilities belonging to firms and individuals servicing the community. In other words, Leaf Rapids has no intention of dying even though located in an environment which is far from hospitable. Significantly, if one thinks of people carrying on after the mine closes down, basically retirees living on savings, public and company pensions, most deserts have charms which are rarely if ever encountered in the rocky forests of northern Canada.

The government decided to make a major change by constructing 600 apartments and 300 houses instead of the reverse. There was bitter disagreement, as the majority of employees wanted their own plot of land. Apparently, they prevailed because, as of 1975, there were 119 townhouses and 74 apartments rented. Also, 196 family units and 67 duplexes were owned by the employees. In addition, there were 35 mobile homes, a fact on which I need not comment further.

Among the terms are a ceiling with municipal taxes of $450 for a single-family unit costing $24,000 and a minimum lot size of 6,000 feet. The cost of land cannot exceed $60 per frontage foot.

A workforce of 80 to 90 percent married people is planned - in effect, a community which is stable and which will grow old in rhythm with the depletion of the mine. It is evident that with such policy and planning, turnover at the Ruttan Mine will be significant, a highly vital factor in this age of increasing technological complexity constantly demanding higher skills.

To complete the picture, the town boasts a school, through twelfth grade, for 625 students which appears to be very well staffed. There is a 40-room hotel complete with all the amenities which make it "a home away from home." With an arena, curling rink, gymnasium, sauna and cultural area composed of theater, library and exhibition center, there is obviously no lack of recreational and intellectual opportunities. The commercial area contains the supermarket, department store, allied shops, coffee shop, bank and business offices. The hospital and municipal offices and council chambers complete the roster of facilities and amenities enjoyed by the community. All of the foregoing are contained within one building of 210,000 square feet which has been the product of a great deal of creative and imaginative architecture. Homes and the town center are placed in settings which respect the need for greenbelts, playing fields and variety - anything to get away from the deadly checkerboard monotony and total lack of character of the company town of the past.

Let us now consider transferring this example to some spot in New Mexico or Arizona. In order to ensure a community which will be viable into the indefinite future, the following steps will be required:

1. Mandatory home ownership by the 80 to 90 percent of the married workforce. Of course, the company would have to provide easy terms - low down payments, modest interest rates and mortgages with a life approximating that envisioned for the mine but not to exceed 30 years.

2. Professional services such as those of the lawyer, doctor and dentist to be obtained by offering facilities for setting up private practices. As the number of such people would be small and collusion regarding price fixing not totally unknown in such

professions, the isolation of most mining towns would dictate that both mine management and town authorities would have to exercise considerable vigilance to prevent rapid enrichment of the few at the expense of many.

3. All retail establishments should be operated and owned by concerns or people with an established track record. Needless to say, contractual arrangements would have to be made which would protect the community and ensure more than acceptable servicing.

4. As the exhaustion of the mine looms, steps must be taken to incorporate smoothly as much as possible of its physical plant into privately owned businesses for supporting the community, i.e., those lathes, grinders, metal working tools and vehicle maintenance facilities which any town needs reserved with payment to be made when the last skip dumps its load at surface. More important, very considerable extensions of land should be prepared for farming. The cessation of mining will release comparatively vast amounts of water, unless the operation has been living at the expense of the water table - something which is happening at an alarming rate in too many places.

5. The community will grow older. Men will reach retirement age. They will own their homes free and clear and have sufficient income to live out their years in comfort and decency. Most will elect to stay right where they are, if where they are happens to be the Promised Land for urbanites in the crowded cities of the rest of the country. Accordingly, sinking funds must be established to constantly finance the creation of new accommodations to house the newcomers who replace those who have drilled off their last round, walked their last beat and settled down to reducing their handicap on the golf course or producing a prize flower.

6. Plans should be based on having the mining company construct everything but purely operational facilities on the basis that the end product will be a self-sufficient town into the foreseeable future and one that will be owned and governed at the earliest possible date by the employees and those who serve them.

7. Finally, when mining operations cease, outsiders should be allowed to purchase homes as they become empty because of death or departure.

Such a scheme will not work everywhere. But the social and economic gains should be sufficiently attractive to provide ample incentive for all concerned to plan from the beginning a continuing life for the community.

11 Small-Scale Mining in the Developing World: Problems and Opportunities*

WHAT IS SMALL SCALE?

Small is very much a relative term. To an ant a beetle is very large, as is an elephant to a man. Similarly, the small-scale mine of Canada, the United States or Australia would appear enormous in most cases to the indigenous operators in countries such as Chile and Bolivia. In the former places five hundred tons a day is definitely considered to be minor output in an age when skyrocketing costs for all of the productive elements - can only be counterbalanced by the economies of scale. When compared with the average nationally owned operation in the developing countries, very little mining which takes place in the industrial lands would be considered as small scale; perhaps some narrow-vein contractors in the western United States, miners extracting uraniferous logs in the Morrison sandstone on the Colorado Plateau, and the handful of people who still pan for gold here and there, would be classified as brethren by most small miners in other parts of the world.

Accordingly, it must be clearly understood that the word "small" as used in this paper means very, very small indeed: instead of five hundred tons a day, more often it means ten tons of hand-cobbed ore - and quite frequently those ten tons will be produced in a month rather than a day.

However, before describing the characteristics of such operations and venturing to make some observations, it could be of interest to look into the size distribution as given in the annual review by Mining Magazine (September 1976), of the mines of the world which produce

---

* Based on a paper prepared for the United Nations Institute for Training and Research (UNITAR), January 1978 and presented at the Small Scale Mining of the World Conference held in Jurica, Mexico, November 26 - December 5, 1978. The copyright is reserved by UNITAR.

150,000 tons per year or more. Evidently, the bottom figure indicates that any operation producing less than 500 tons per day has to be classified as a small mine by international standards - standards set, of course, by the industrial world. But, as noted above, such scale would be considered gigantic in more than a few developing countires. In any case, Mining Magazine (September 1976) lists 1,126 mines. This does not include mines of the centrally planned economies for which data are evidently not obtainable or, at least, not in sufficient detail to be compatible. Fossil fuel mining and production of such commodities as building stone, sand, gravel, limestone, etc. are not part of the tabulation. The latter group - the construction materials - are of very great importance. In fact, because of the enormous tonnages involved, their annual value of production exceeds that of the entire metallic family. However, statistical data are fragmentary, so that inclusion in the listing is not practicable and could be misleading.

In terms of numbers, of the 1,126 mines 40 percent are located in only four countries, all of which are industrialized - the United States, Canada, Australia and South Africa - and the majority of tonnage is also produced in those four countries. In view of the fact that the combined area of these nations is only about a quarter of that of the developing world, it is evident that deficiencies in regard to availability of capital, infrastructure and know-how condemn most of the developing lands to underexploitation of natural resources. In such environments, small-scale mining has heightened importance. Too often this is a most unfortunate situation as I hope to show.

GENERAL CHARACTERISTICS OF SMALL-SCALE MINING

The remarks that follow apply particularly to the developing world.

1. The industry is important. Frank Skelding (1972), in writing an authoritative study for the United Nations estimated that about ten percent of the world mineral production is provided by small miners.

2. Mining Magazine calculates that outside of centrally planned economies there are more than 6,000 mines which produce less than 150,000 tons per annum. It is almost certain that this estimate is very much on the low side.

3. Small-scale mining is a labor-intensive industry, usually of an artisanal nature. Too often it involves nothing more than the direct application of human energy: e.g., a pick and shovel, without greatly multiplying that energy by mechanical means such as a rotary drill and loader.

4. It is also a brutalizing business. Labor is almost invariably exploited. Submarginal wages are paid, and working conditions are too frequently appalling. I will never forget the tortured faces of miners in the Andes, with 50 kilos of rock on their backs struggling up steep

inclines with vertical rises of as much as 200 meters, and the families of workers living in the mouths of abandoned adits. Goats were better housed and, in terms of value, probably rated higher.

5. The labor force is usually unorganized and operates outside of any social legislation which may be on the books. The larger mining companies will be held responsible for paying a decent wage and for provision of schools, hospitals, recreational facilities, and sound living conditions. In addition, sinking funds will be set up to meet pensions; medical insurance will be provided; and various bonuses will be mandatory. The small mine owner, in contrast, will usually accept only one obligation - namely, to pay a miserable pittance with reasonable promptness.

6. Similarly, there is almost always total disregard of laws concerning health and safety. A considerable amount of dry drilling still goes on in various parts of the world. I can very vividly remember my first descent into a mine in Chile shortly after joining the United Nations. The shaft was vertical and over 200 meters deep. The cage consisted of a piece of hard rubber about the size and shape of a bicycle tire which was connected to a small, 1890s vintage hoist by a quarter-inch steel wire rope red with rust. In places where safety procedures are enforced, a quarter-inch cable can be used only for hoisting materials, and replacement is so frequent that there is no chance for rust to form. Not surprisingly, the rope described above broke about two weeks later, dropping two men to their deaths. Another cage which is often encountered consists simply of an oil drum. The mere act of getting in and out of such a contraption has decidedly dangerous elements. In effect, too often life is cheaper than equipment. To properly equip even a small shaft will cost many thousands. To dispose of a body, probably less than a hundred dollars will be required. And if the bad air leads to silicosis - in the end it almost always does - when the miner's usefulness becomes impaired, he will simply be let go and there will be no consideration given to compensation.

7. As noted earlier, the small-scale mining industry is of course very labor intensive. The human muscle is too often the dominant force. As I shall discuss later, when reviewing the role of governments in regulating and, on occasion, stimulating the sector, this can produce dilemmas of no mean proportions.

8. Partly because of antiquated laws in regard to the very limited extent of permitted mineral holdings, and partly because the nature of operations does not require much space, small mines tend to be very small indeed in a real sense. Accordingly, a desolating characteristic, too frequently destroying what could have been and should have been important mines, is the fractionation of ore bodies. A deposit containing, let us say, ten million tons of highly exploitable ore - an important national asset - will frequently have as many as 40 different owners, each with their few hectares of property, each without even a distant approximation to the resources, either technical, managerial and/or financial, required to undertake meaningful and rational exploi-

tation. Whereas two or three shafts would suffice for a mine operated as a unit, 20 or 30 "gopher holes" are pushed in or down without any engineering, any overall geological knowledge of coordination, designed only to extract a few tens of tons daily - if that. And once a multitude of owners blanket a mineral deposit, there is no way to bring all parties together to obtain a single direction and rational exploitation as a unit except by expropriation by the government. This happens rarely if ever. The politician and public authority, if in a rapacious mood, will zero in on the haves rather than the have-nots. Chucquicamatas and Siglo Veintes are considerably more attractive than deposits which have been more or less raped, the potential largely lost and with which existing workings do not constitute assets but instead troublesome barriers to efficient mining operations.

There are several essential features of good mining procedures:

a. Sound knowledge of the location, attitude and composition of ore zones is necessary. To obtain these data, diamond drilling and drives - drifts, cross-cuts and raises - are the standard exploratory media in more favored parts of the world. They are practically unknown where small mining operations are concerned in the developing countries. Instead, the norm is the discovery of a rich vein or lens, always at surface because the industry does not have the technical capacity or money to delineate buried deposits. Then, the mineralization is followed with extraction paying for further progress. That to one side or another there might be a richer vein will rarely be a matter to be investigated. "The bird in the hand" controls.

b. There must be strict respect for the horizontal plane with development headings such as drifts and cross-cuts (with, of course, allowance for about a degree of inclination in favor of loaded cars). In addition, it is recognized that with shafts - either vertical or inclined - the straight line is more or less mandatory. To achieve these desiderata, the theodolite is a necessity. Sadly, in the small mines of the world, the mine surveyor is very rarely encountered. Accordingly, workings head in all directions and on all planes, and too frequently there is no way of relating one to the other.

c. Direction and control by mining engineers and geologists is mandatory. Such personnel are too expensive to be employed by the vast majority of small mines. However, it has to be conceded that in places with a long-established mining tradition, a breed very skilled in "sniffing out" where the richest ore lies always exists.

A considerable number of the mines which produce 20 tons per day or less could, under more favorable circumstances, achieve listing in Mining Magazine. Excluding alluvial workings, by and large small mines operate on vein-type deposits. But frequently there is not a total cutoff in values between veins. For example, separating the massive chalcopyrite or lead-zinc-silver veins running more or less parallel and fairly closely spaced, there will often be wall rock with disseminated mineralization. The whole may have very respectable grades with dimensions which could support medium- to large-scale mining. But lack of knowledge of the characteristics of the ore body and technical capacity, together with the mercenary intent of making as much as possible, as fast as possible with the least investment possible, leads to ruthless high-grading. Instead of mining over widths of, let us say, ten meters, 30-centimeter veins are gouged out, the cream is extracted, leaving nothing but skim milk. It is to be hoped that future generations may be able to exploit the material left behind. But this is by no means certain in too many cases.

Much small mining is socially beneficial in that it is seasonal. The crop is harvested and the winter lies ahead. Men then head for the mines. Larger scale operations cannot accept such transitory labor; the small operators, who have no appreciable capital tied up, frequently can.

Earlier, I pointed out that the industry's special characteristics lead to great wastage more than occasionally. On the other hand, it must be recognized that a large number of mineral deposits can only be exploited on a miniscule scale. Very few concerns are interested in a reserve of a few tens of thousands of tons of fluorspar, barite, or even high-grade base metals. In fact, more than a few of the small miners have gone into business by taking the leavings of the large mining companies. The local entrepreneur who seizes the opportunity will often become a wealthy man.

GOVERNMENT POLICY

The industry will usually require special treatment from government because its structure is almost always weak. Therefore, it becomes essential to define what is meant by small-scale mining. This is not easy. Most definitions tend to be based on quantitative factors. Often this is too simplistic an approach because of the exceedingly numerous permutations and combinations introduced by the special characteristics of the mineral itself, by grade, physical setting, and a host of other factors. It will have been noted that Mining Magazine, in effect, classifies any mine producing 150,000 tons per year as being important. With relatively high-grade gold, lead, zinc, copper and such, this would be so. With the beach sands being mined for rutile in Australia, as a rule of thumb, it has been found that a throughput of less than three million tons per annum is not economic. Similar, if not greater, size demands are made with porphyry copper deposits and low-grade iron ore for conversion to pellets. Other ways of defining what is small involve the degree of mechanization or lack thereof, as expressed by output per man-shift; capital investment per unit of output or per productive worker; or a combination of these and other criteria. I suggest that governments should classify as "small," and accordingly give special treatment to, operations which have very limited potential in regard to reserves, are undercapitalized, produce only minor amounts of mineral, lack technical direction and sound managerial inputs, and waste resources, both physical and human.

Obviously, governments should devise systems for acquiring an understanding of the place of small mining in the national economy. A certain amount of statistical measurement is imperative if there is to be any control over the industry. While such measurement can only occasionally be precise, there should be a reasonably close idea of what is produced quantitatively and qualitatively and where; the composition of the labor force; capital investment, if any; commercialization of output; and profits and losses. In effect, the small miner should, in most cases, be forced by law to maintain accounts. Too frequently all bookkeeping is done in the owner's head and almost always in such ways as to ensure minimum tribute to taxing authorities. However, in imposing some approach to sound business procedures, rules and regulations should be clear and simple. The accounting and reporting requirements that a medium or large mining operation would routinely satisfy could not be supported by the vast majority of small miners. Indeed, one class would have to be exempted from any written communication with government - the artisanal miner digging and panning a small amount of alluvial ground for gold or diamonds or handpicking through an old waste dump. Such people will usually be illiterate, and their annual income will seldom suffice to pay the monthly wage of a competent accountant.

In the same vein, health and safety measures should be enforced. But again regulations would have to be applied less stringently than with large and well-financed concerns. Nevertheless, minimum standards can be enforced which will not put the mine owner out of business. Hard hats and safety boots are not costly. Ladders are cheap to maintain. And nowhere in this age can there be any excuse for permitting dry drilling underground.

As stated earlier, the small miners can make appreciable contributions to a nation's economy. Accordingly, it will frequently be advisable to extend public assistance to the industry, chiefly by strengthening the infrastructure - roads, power and telephone lines, etc. Often, far smaller contributions can have a major impact. I know of an instance where the introduction of sluice boxes costing only a few hundred dollars increased primary recovery with a gold placer from 50 percent to over 90 percent. It is a sad commentary that these boxes were quickly destroyed and the United Nations team, which masterminded the scheme, was sent fleeing into the hills. The socioeconomic organization involved was a model of simplicity: the men, the miners, washed the ore they extracted over rocks but, in the process, recovered only half of the gold. A more inefficient system would be hard to devise, but every bit of the yellow metal belonged to the miner. If his wife and children wanted to eat, they did so by the proceeds from panning the overflow downstream. With the introduction of efficient sluicing, malnutrition became such a real threat that the boxes met a violent end. This is a rather extreme illustration of the need to analyze the mores and practices of the community concerned before introducing new technology. Nevertheless, the fact remains that reversion to traditional practice meant reducing recovery from more than 90 percent to, perhaps, 65 percent. Fine flakes of gold can travel a long way downstream in rivers roaring out of the mountains. The dispersion train is impressive. It is therefore advisable to capture the metal as close to the source as possible.

Other types of aid which should be provided are low-cost mineral analyses, beneficiation testing, geological mapping and the provision of technical advice, topographic surveying, etc. Such assistance will usually increase efficiency several fold.

Particularly with metals, small-scale miners frequently produce high-grade hand-picked mineral, not concentrates. This is not a readily marketable product and usually not one which can bear the cost of shipment involved in exportation. In any case, unless organized in cooperatives which are rather rare entities in an industry operated by highly individualistic people, the miners are frequently incapable of marketing, except through the government. The latter has to become an ore buyer and treater so that the end product is upgraded to at least concentrate form. As exemplified by the record of the Empresa Nacional de Mineria (ENAMI) in Chile, such schemes can work and work well. But implicit in the process is the payment of fair prices and the creation of treatment facilities of sufficient size and efficiency to ensure that the largest part of value accrues to the mine owner. Depending upon the characteristics of the ore body, these people occasionally become rich, but more often it is a marginal enterprise, sometimes so much so that the government may actually have to subsidize to get the feed required to maintain the throughput needed for effective operation of treatment plants.

A parallel mechanism is the mining bank, the principal function of which is to provide credits to the mine operators for purchasing supplies and equipment. Unfortunately, in an imperfect world and one where the civil servant is chronically underpaid in many countries, the mining bank contains the seeds of its own destruction. Where payoffs are very much a part of the system, not only is aid diluted but also debts tend to be uncollectible and supervision of application of loans becomes very lax indeed.

Environmental considerations are very fashionable these days, and they should be - in the industrial world. In developing lands, however, I strongly suggest that governments should not be preoccupied with any pollution which might stem from small-scale mines. It has to be negligible because the industry is a very minor consumer of fossil energy. The strong backs, legs and arms are the principal producers of foot pounds and human sweat has no ecological impact.

With very high unit value minerals such as gemstones, gold and platinum, governments very frequently try to buy cheaply. This is a mistake. Such policy largely enriches the illicit buyers who ensure that their purchases do not figure anywhere in national accounts but instead enter openly into commerce somewhere abroad. A wise government will pay the top dollar.

The industry usually requires a special tax regime. Even if sufficient control exists to obtain sound data on profits, a graduated scale of income tax should exist because earnings will tend to be very much on the scanty side. Usually, however, most small-scale miners will not produce the data required to assess taxable profits, either through inability or design. The government will then be inclined to impose ad valorem taxes assessed at the ore-buying stage. In such cases, a great deal of moderation is strongly recommended. Even with efficient, well-financed mining operations, net per dollar of sales will average out around 15 cents before taxes; it will not be higher in most cases with small, primitive workings. Of course, this assertion will be countered by some on the basis of the low cost of labor. Actually, such workers are usually seen to be anything but cheap when meager wages are related to even more meager output. In any case, when considering ad valorem taxes it should be realized that, in general, a one percent levy on the gross value of production will be about equivalent to seven percent of net profit.

The developing nations quite rightly wish to achieve status equal to that enjoyed by the industrial world. That it is highly doubtful that many will be able to do so is not a subject to be pursued in this chapter. The movement quite probably will be in the other direction as growth slows in the developed nations. In any case, the desire to modernize, to obtain considerably greater benefits from the technology of the age, will present more than a few governments with a problem of very severe proportions. A number of years ago, I undertook a three-month consulting job in India concerned with working up a development plan for a large iron ore deposit based on the very latest technology and the maximum degree of mechanization. Part of the final product was an estimate of operating costs. When the report was presented to the official who had commissioned the work, his first question had to do with their order of magnitude. When told what they would be, he wanted to know how they compared with those of a nearby mine on the same iron ore range which was being worked almost entirely by hand labor. He was, of course, shocked when informed that they would probably be more than three times higher, that tire wear and maintenance of vehicles alone would exceed the total operating costs of the unmechanized mine. The answer is very simple: modern technology cannot compete costwise when laborers earn, as in this case, between 15 and 20 cents daily. Accordingly, when governments view the small mining industry and decide to pour in money to upgrade operations, there will have to be serious consideration given to the effects on employment. As noted before, the industry usually has a brutalizing impact on the worker, and almost always his gain is very small. There is a tragic dilemma here: if mechanization is undertaken, the urban slums fill up; if operations remain primitive, the employed are condemned to a miserable existence. To find a midpoint, a workable solution, would test the wisdom of Solomon.

Finally, archaic legislation and/or almost total failure to force responsibility upon those who hold mineral rights condemn mineral development to virtual stagnation in many countries. Concessions are obtained without ever putting foot on the property simply by drawing some lines on a map in a registry office and paying a nominal fee. Annual taxes may be laughable or, better said, tragic. In Chile, about 20 years ago the rental, unchanged since 1895, amounted to the cost of two cigarettes per acre. And cigarettes were very cheap. Where regulations state that mineral rights can only be maintained by virtue of continuing investment in exploration and development of the property, there is far more observance of the breach than the promise.

Governments simply do not enforce the laws or regulations in many countries because of indifference, lack of personnel for inspections, fear of antagonizing a substantial body of voters, and many other considerations and factors. The principal result is that ground is tied up, seemingly in perpetuity, by speculators, or by people who have virtually no resources to define and exploit any minerals which may lie within the property. It would be a mistake to assume that such impoverished claim-holders are easy to deal with, that the flashing of a roll of currency will be responded to eagerly. Instead, the sight of money generally whets appetities, often to a degree so ridiculous that no deal is made and the property concerned continues to lie fallow, totally neglected, contributing nothing whatsoever to the owner or the economy. Accordingly, I strongly believe that acquisition of mineral rights must involve staking on the ground rather than on paper, to be followed within a reasonable time by precise delimitation by a licensed surveyor and, most important, strict enforcement of annual work requirements. These should be sufficiently exigent to force sustained effort to determine the mineral potential of the property or abandonment.

CONCLUSIONS

The small-scale mining industry is highly inefficient, exploitive in terms of human endeavor and generally wasteful of natural resources. But it produces a great deal of wealth and, as such, is very important in many developing economies. It is a fertile source of employment, and it extracts minerals which would lie untouched if left to the devices of larger, better organized mining concerns, either public or private. There are other pros and cons. In the final analysis, small-scale mining will always be with us. The problems in the future will revolve around steps to be taken to humanize, rationalize and optimize the contribution that the industry can make. These steps will have to be a function of public action if they are to be effective.

12 Epilogue

The preceding chapters were written between 1965 and 1978. How much has time eroded, altered and even rendered invalid the ideas and observations in them? Had the world stood still since the earliest was written - and had my mind been congealed ever since - the answer would be simple. Instead, the years over which these papers were written have been perhaps the most terrifying of what can be described as the most complex and dynamic of all centuries. Change - political, economic, social and institutional - has been the order of things. Besides, only the zealot has a single, burning beacon; the man caught up in the fast swirling and ever-changing currents of a turbulent time will hear a dozen different siren songs. Views will be revised, theories may or may not stand up, and judgment based solely on distant experience will lead to dead ends as the world changes. Accordingly, it may be of interest to examine the evolution of what I considered to be "verities" at one time or another. I shall concentrate, of course, on mineral resources because only from one's field of competence can illustrations be provided which bear on broader issues.

> Acceptance that progress will remain with us as the basic fact of life, progress being defined solely in materialistic terms.

This proposition today seems to have been more reflexive than reasoned and was based largely on observation of the American model. Even a casual look around reveals that the essential dynamic of today's society is exaggerated consumption, i.e., the production of ever more goods to satisfy tastes already sated in too many instances. How then can progress cease, how can brakes be applied? Let us consider this in relation to the automobile. As vast quantities of nonrenewable resources are poured into a product which befouls the atmosphere and engenders economic crises, public transportation limps along in a state of near bankruptcy and the finest agricultural lands disappear under pavement. Roads tend to follow rivers rather than corkscrewing from peak to peak. Rich bottom lands are gobbled up. It has been estimated that the highway system of the United States covers almost three percent of the total land area of the country. What would follow an edict limiting each family to one car? In a world where most people walk, or at best cycle, this would not seem to be deprivation. But the enforcement of such an act, unless spread over many years, would tear the nation apart. Unemployment on a totally unacceptable scale would ensue. It would mark the end of the "dream," the spiritual destruction of a people. They would not go down without fighting, and they would almost certainly take the rest of the world with them.

Similarly, the industrial-military complex is, in effect, the whole of American industry. Few of the truly important companies could break even without revenues from the sale of military hardware on a very large scale. Curtailment of the space program, something which has decided defense implications, would strike a mighty blow to many. Without contracts from the Pentagon, far too many research centers and consulting firms would go out of business. Of course, it is not only the United States which is so afflicted: the manufacture of armaments is a primary mover of many economies of the industrial world.

Progress, defined as sustained growth, is evidently a sine qua non in the mind of twentieth century man. But what does sustained growth mean in terms of the minerals required to support it? Let us consider copper for illustration. Currently, the world consumes about eight million tons annually. In recent decades, consumption has been increasing at a rate of about four percent per annum. If this rate were or could be maintained over the next 25 years, annual world requirements would reach about three times their present level. An average of about 600,000 tons of new copper would have to come into production each year, that is, if no mines, new or old, closed down. The sheer magnitude of the effort necessary to find the required reserves which, by definition, lie at ever greater depths in ever more remote places, is difficult to fathom. The energy inputs required to extract the vast tonnages would also be enormous in quantity and cost. With everything - financing, construction, operations - scales would be so gargantuan as to be impossible to achieve in many instances and highly undesirable in most cases.

What of the concept of sharing more equitably in global progress and wealth? Implicit in that concept is the hope that the developing world can come up to the levels enjoyed by the Western world rather than movement in the opposite direction. With regard to per capita consumption of basic raw materials, it would not be difficult to demonstrate that endowing all with the equivalent of that enjoyed by the average American citizen would involve astronomical quantities beyond any conceivable productive capacity, not to mention absorptive capacity! Besides, as the world is now constituted, expansion of its physical plant would tend to have an adverse impact upon the air, water and land the total environment - roughly proportional to the amount of macrogrowth involved. The ability of the world to handle poisons, noxious fumes and other destructive wastes is decidedly finite, and already too many limits are being severely strained.

If the world economy makes a significant recovery, governments will exert every effort to maintain momentum regardless of longer term consequences. The juggernaut which is all but out of control now will roll on. What will be needed to stop it is quite basic: new wisdom and vision; greatly heightened awareness that all men are really brothers in that the same threats to survival apply to each and every one; and concerted action, with the rich nations ready to make more than a little sacrifice and the poor willing to scale down their aspirations. Only a fool who has not read history could be optimistic that such radical change in the character of man can come to pass, that the new mentality required will spring forth in time.

Belief that under sane management the earth is so fruitful that it can supply for all time the mineral needs of a world population in balance with food supplies.

Nothing stated in the previous section should in any sense be considered as contradicting the observations made in Chapter 7, "Are We Running Out of Minerals?" The problem has little to do with reserves but instead revolves about the management of these in particular, and the state of the world in general. Also, implicit in the phrase "a population in balance with food supplies" is the urgent need to arrest the growth of the people living on this planet. On the one hand, the majority suffers from malnutrition; on the other, there is a decidedly low ceiling on the possibility for harvesting substantial new sources of food. Good land is limited and so is fresh water. In fact, of recent years probably more acreage by far has been reduced to desert than brought under cultivation.

Belief that in most developing countries mineral resource development offers the best opportunities for significant change.

As the years have passed, it has become evident to me that the "wondrous benefits" of mining developments can, in many cases, stand re-examination for several reasons. First, it is a stark, sad fact of life that a significant number of countries have only negligible mineral potential. Second, mineral developments of importance require a very long gestation period, often ten to fifteen years. A "teaser"-type situation often results: lots of indications of mineralization that could be economic, lots of hope and yet nothing concrete enough to lay a proposition before the bankers - all of these may be examined and explored a half dozen times over many years before a final answer, quite often negative, is obtained. Third, mineral exploration is highly costly and becoming more so with every passing day. Each year the search has to be carried out in more remote places or at greater depth, or both. Prudent management today should not earmark less than $7.5 million for a five-year exploration program starting in an area of considerable size where the general geological picture is known. Even then chances will not be good for turning up more than an ore body which will support medium-scale operations. Finally, in most of the developing world, an important mine will usually contribute more to outsiders than to the nation itself. Modern mining is not labor intensive. The "downstream" impact tends to be negligible where local industry is only slightly developed. The largest part of revenues will go to service external debt, employ foreign technicians, import supplies and equipment, pay marketing costs, and return a profit to the foreign investor.

In view of the foregoing, sound development priorities should be based on comparative cost/benefit studies of mineral exploitation and the production from the farm, the sea and the lumber mill. After all, agriculture is the basic industry in most developing countries, the majority of which will never be industrialized. Not only do they lack the required ingredients, but it is also too late, if one accepts the thesis that material growth has peaked or will do so shortly.

> Conviction that governments, particularly but by no means exclusively those of developing nations, have no real conception of the risks and difficulties involved in seeking and exploiting minerals and therefore are usually incapable of creating a workable investment climate.

This is not a black or white situation. The inability to see beyond one's nose can afflect policy making in developed countries as well, as I have demonstrated in Chapter 9, "Evolution of Mining Taxation Policy in Canada." I have expressed over the years strong views on the dos and don'ts for creating a sound investment climate. I wish to warn here particularly against falling prey to merchants of gimmickry. This is why I say, "Keep the lawyers in their place." Too often they are the tail which wags the dog. A few years ago I had a talk with a major Wall Street underwriter of bonds. He mentioned that the vast majority of deals are worked out verbally and sealed with a handshake. The principals meet and, with little fuss, reach agreement on the essentials. Often this is done over lunch. Later, the lawyers take months to tidy up the details, but they are only peripherally involved in working out the grand design. In effect, lawyers are rarely qualified to be the architects but instead should fill the role of draftsmen. Admittedly, in countries short of technical talent but inundated with legal lights, it may be difficult to keep lawyers from the center of the stage.

It is not advisable to be too clever when drawing up legal documents such as mining codes. For example, it is very much the mode in the developing world today not to concede exploitation rights when negotiating exploration concessions. In the hope of achieving this objective, some very fancy semantic exercises are resorted to. Phraseology is used, such as, "The concession holder shall have the first right (or priority or preferred option) in regard to the exploitation concession." These meaningless words, which have no force in law, simply insult the intelligence of the potential investor and give him a clear signal that the government has no intention of acting in good faith. Another favorite ploy is to have every right so qualified - with the likes of "without prejudice to the rights of the state," and "in conformity with the law" - as not to constitute a right.

> Despair that the developing countries can and will enact mining and fiscal codes which reflect the realities of the age and, therefore, will attract private mining capital.

By and large, the despair lingers. From time to time a new mining code comes upon the scene but with all the impact of a feather floating onto a bubble bath. In the first place, most mining codes are characterized by a jungle of safeguards, supposedly of the national interest. Unfortunately, the safeguard of one man can be the stick of dynamite threatening the other. Second, they tend to be peas from the same pod. In fact, some people have made a profitable business of wandering hither and yon advising on form and content of codes. Sadly, they offer the same version everywhere. This is a gross error: a mining code must be very carefully tailored. It should never be a standardized document; instead, it must reflect the distinctive character of the country concerned, namely, the political system, philosophy, aspirations, physical setting, data base, financial resources, state of development, capacity, skills and attitudes of the people - from the man with a shovel to the top rank. All of these must be considered and weighed, frequently intuitively, and fed into the equation for establishing the parameters for development decisions and the framework for such decisions, namely, the mining code.

In this volume, I have suggested guidelines for drafting a mining legislation which, while protecting the national interest, does not scare away development capital. I should stress that it is equally important to avoid granting excessively liberal concessions such as lengthy tax holidays, particularly where these do no more than enrich the treasury of the country of origin of the investment capital. Actually, today, the sophisticated mining house itself will usually not care to negotiate for unreasonably favorable terms. By now, there is widespread realization that over the long period required to bring a major operation into production, the original contract is bound to be re-examined two or three times by succeeding governments and drastically altered if there appears to be too much of a free ride.

> Certitude that skilled planners capable of looking a long way ahead are vital precursors of development.

This belief has been rudely shaken, as it has steadily become more than obvious that problems are too immediate almost everywhere and crystal balls grow ever more clouded. In fact, a good case can be made for a "live for today" approach in many parts of the world. Long-range planning every day becomes more of a futile exercise. There is simply too much turmoil, too many uncertainties, too many unknowns and variables. The United States should have a highly developed capability for looking into the future. It has the experienced people, has developed the concepts and tools, and can feed vast amounts of data into the most sophisticated computers. Yet, I can recall two forecasts made by competent public agencies of the needs of the United States in the year 2000 for the more common and essential minerals. The forecasts varied by factors of as much as seven. Mention might also be made of the "bomb" which the Organization of Petroleum Exporting Countries exploded late in 1973 - a bomb which should have been expected but was not, except possibly in a few board rooms.

The most pressing reason for taking action today in the developing world is that problems are so multitudinous and immediate that they simply cannot be brushed under the carpet. And, as stated earlier in the book, the concern of governments has to be with those living and suffering now rather than with the yet unborn. If the welfare of the former is improved intelligently, future generations will come aboard a tauter ship.

Attraction to exploration for "elephants."

The United Nations has established a unique record in regard to exploration for the prominent minerals, those most involved in international trade. In retrospect, I think that there should have been a slight shift in emphasis. Instead of always dreaming about bonanzas such as Chucquicamatas, Sudbury Basins and Katangas, it probably would have been advisable to have built up the capacity to seek more mundane things. Geological settings, of course, have to be paramount; they determine the focus of search. But few of any size do not contain ceramic clays, limestone, building stone, aggregates and other neglected industrial minerals and rocks. Such materials can usually be exploited by local interests for the national market with perhaps some overflow across neighboring frontiers. Thus, most of the fruits will be consumed at home and many will have a share.

The conviction that nationalism is the sorriest affliction that can beset man.

As the years pass, my detestation and fear of the drum beaters, the proponents of the Thousand Year Reich and Manifest Destiny, those who proudly paraphrase "my country right or wrong but still my country" has not abated one little bit. Instead, the movement has been in the other direction.

It is up to the reader to sum and balance what is in these closing pages. I am aware of one thing: I have sung a mournful song, and I am glad it is finished. So now I think I will go to an Irish pub, sip a beer or two and, while listening to God's anointed tellers of tales, remember Walter Hagen* saying, "You pass this way only once, so stop awhile and smell the flowers."

---

* An American golfer of immortal skills and a bon vivant of most impressive stature.

Appendix

INTRODUCTION

1. The United Nations Revolving Fund for Natural Resources Exploration (the Fund) has been created, as a separate unit with its own identity, in order to extend and intensify the activities of the United Nations system in the field of natural resources exploration in developing countries, utilizing for this purpose voluntary contributions and funds generated through the production of resources discovered or developed with the assistance of the Fund in such a manner as to ensure its revolving nature derived from the self-help principles for the mutual benefit of developing countries. (1)

2. The Fund, established as a trust fund by the General Assembly and placed in the charge of the Secretary-General, is administered on his behalf by the Administrator of UNDP. The Governing Council of UNDP will be the governing body of the Fund for the first four years at the end of which, according to Economic and Social Council resolution 1762 (LIV), the Economic and Social Council will review, in the light of the experience gained, the functions and institutional arrangements as well as the repayment system of the Fund, with a view to recommending to the General Assembly changes and improvements, taking fully into account the comments of the governing body and the Committee on Natural Resources.

3. In General Assembly resolution 3167 (XXVIII), the General Assembly requested the Secretary-General, in close collaboration with the Administrator of UNDP and with the participation of the International Bank for Reconstruction and Development (World Bank), to prepare operational procedures and administrative arrangements of the Fund for approval by the Governing Council. The Secretary-General accordingly presented a report to this effect to the eighteenth session of the Governing Council (DP/53) and subsequently issued a revision (DP/53/

Rev.1) to take into account the recommendations of the group of mining experts convened by the Administrator in April 1975 (DP/129). At its twentieth session, the Governing Council adopted certain decisions concerning the Revolving Fund, approved the operational procedures and administrative arrangements contained in DP/53/Rev.1, insofar as they were consistent with these decisions, and authorized the Fund to start operations (E/5703/Rev.1, paragraph 536).

4. The following operational procedures and administrative arrangements of the Fund reflect the decision of the Governing Council at its twentieth session.

I. OPERATIONS

5. The operational procedures and administrative arrangements set out below are based on the following premises:

(a) The Fund shall provide a form of cooperation in the field of natural resources to developing countries additional to the assistance available to these countries through projects financed under their country programmes within their indicative planning figures (IPFs), or through regional or interregional projects when appropriate;

(b) While the Fund can ultimately become self-sustaining, it must until then rely upon voluntary contributions;

(c) When the revenues of the Fund eventually become sufficient for it to be self-sustaining, the Fund shall provide a means for mutual assistance and co-operation among developing countries;

(d) As a new and unique venture, the Fund's procedures and arrangements must be flexible and experimental, especially as the Fund will be largely devoted to exploration activities which inevitably involve a high element of risk and uncertainty, and its success will be dependent upon a prudent approach to project selection, good management in project execution and the employment of the best available professional expertise.

6. The Fund will build up its own technical and managerial staff as expeditiously as possible and commensurate with its resources and the volume of operations. In the initial stages it will utilize to the greatest extent possible existing staff in the United Nations system, thereby keeping the Fund's overhead costs to a minimum. Any additional costs incurred by United Nations agencies in respect of activities on behalf of the Fund will be reimbursed to the respective agencies out of the Fund's resources.

7. The Fund is distinguished from other forms of development assistance provided by the United Nations system by two factors: (a) the revolving nature of the Fund and (b) the fact that Replenishment Contributions will be based on the proceeds of production following a successful exploration project.

Operational Principles

8. The revolving nature of the Fund affects the operational principles with regard to (a) the type of natural resources to be explored, (b) the criteria for project selection, (c) the types of services to be provided and (d) the phased approach to the design of projects.

9. All types of natural resources exploration may, in principle, be included within the scope of the Fund. The natural resources to be explored should have a potential market, whether domestic or external, and the exploration activities required for their discovery should fall within the financial resources of the Fund. Owing mainly to the fact that certain types of exploration would probably be too expensive to be undertaken by the Fund until it has received or earned substantially larger resources than are likely to be at its disposal during the early stages, it is believed that the Fund should concentrate its activities during the initial stages on exploration for solid minerals.

10. Projects shall be selected on the basis of their technical and economic viability and, therefore, their potential for contributing to the requesting country's economic development and ability to help achieve the revolving nature of the Fund. At the same time, however, in selecting projects due consideration shall be given to the equitable distribution of the Fund's resources and to the special situation of the least developed countries and of developing landlocked and island countries, consistent with the revolving character and objectives of the Fund.

11. The services to be provided by the Fund shall be exclusively in the exploration and associated pre-investment fields, broadly defined. They shall not include broad geological surveys, training or institution building, as such, although they will naturally provide opportunities for on-the-job training of nationals, transfer of technology and utilization of national institutional facilities to the extent that these contribute to the expeditious execution of the project as determined by the management of the Fund. Such broad surveys, training and institution-building projects, as well as the other technical assistance and pre-investment services traditionally provided by UNDP, shall continue to be available to the developing countries under their IPFs.

12. In order to ensure effective UNDP support for the development of the natural resources sector in user countries, maximum efforts shall be made to co-ordinate projects financed by the Revolving Fund with UNDP's current and future institution building, institution strengthening and manpower training programmes in this sector.

13. The Fund may engage in such research activities as would contribute to the fulfilment of its main objectives, consistent with the revolving character of the Fund.

14. Project operations shall normally be designed in phases. This will enable the Fund, throughout an exploration sequence, to take the appropriate decisions to modify or re-orient project activities and to

discontinue less promising avenues to avoid losses. By phasing and selective progression, the Fund's projects shall concentrate on those areas and activities that, in the light of intermediate project results, are considered most promising. Exploration activities will generally fall within the following categories and may be initiated or terminated at any phase of the exploration sequence:

(a) Preliminary activities to verify and evaluate geological and related data provided by the applicant country (2) and limited field checking within the area or areas said to possess good potential for mineral development;

(b) Technical reconnaissance within the chosen area or areas using photogeological, geochemical, geophysical and other techniques to define more precisely targets for detailed prospecting;

(d) Detailed evaluation including large-scale mapping of the targets, detailed geochemical and geophysical investigation, trenching, pitting and prospect drilling to limited depths to outline geological structure in depth, and as far as possible to obtain information on the order of magnitude of tonnages and grades of possible ore bodies. Additionally, limited beneficiation tests and preliminary studies of infrastructure, marketing, profitability, etc. may be included.

15. The successful conclusion of 14(c) can be defined in regard to most mineral deposits as completion of the pre-feasibility phases, e.g., indication of a deposit which is potentially economic. Further work required to produce the data which, when and if viability is established, will provide the basis for bringing together the financial and technical requirements for reaching the production stage is defined for Revolving Fund purposes as the feasibility study. The data referred to above will result in relatively precise definition of tonnages, grades, principal minerals and by-products, beneficiation methods and recoveries, infrastructure and manpower needs. and all the economic elements which will determine the profitability of the enterprise. Such investigations require heavy expenditure closely related to the production stage of resources development. Expenditure of this magnitude is likely to be beyond the Fund's financial capacity in the initial years, as Replenishment Contributions will be limited as indicated in paragraphs 20, 21 and 22 below for the earlier phases of exploration defined in paragraph 14. The Fund could, however, undertake feasibility investigations at the request of Governments as soon as it is financially in a position to do so; in these cases special arrangements for reimbursing the Fund for these costs would be required.

16. An exploration programme may commence with pre-project activities and may continue through subsequent phases as favourable results are obtained. Where earlier exploration, whether financed by UNDP or other public or private souces, has yielded favourable results, the Fund's activities may commence at a later phase in the exploration sequence.

17. Phasing and selective progression, whereby decisions on further activities are dependent on technical and economic assessment of

previous results, requires an objective decision-making capability by the Fund as well as authority to terminate a project when findings are disappointing or other circumstances impair its chances of success. As the ability to continue, terminate or modify a project is basic to the effectiveness of the Fund, the Project Agreement with the user Government (see Chapter III) shall stipulate such authority. The Fund shall take such decision after consultation with the user Government. Moreover, in order to assure user Governments that the Fund will proceed vigorously, the Fund will be obligated to carry out at least a minimum amount of work as the first phase of a Fund project. When the exploration results of a Fund project have attracted public or private investment interests of a quality and quantity assuring effective achievement of objectives, the Fund should be in a position to discontinue its activities while maintaining its right to Replenishment Contributions.

Replenishment Contributions

18. Replenishment Contributions to the Fund are not intended to be repayment of loans but rather to provide the Fund, from its successful projects, with a share of the proceeds from resulting production expected eventually to make the Fund revolve, so that its resources will be replenished and the Fund will be able to continue its exploration activities for the benefit of developing countries. Governments requesting projects shall be required to assume an obligation to make such Replenishment Contributions to the Fund (a) only for successful projects when the resulting mines come into production and (b) at rates based on the value of the resources produced, limited in time. These obligations shall be set forth in a Project Agreement which shall be effective for a period of 30 years.

19. In order to identify the successful projects that may lead to Replenishment Contributions to the Fund, the Project Agreements for exploration activities between a Government and the Fund shall delineate and define the project area to be explored. The project area shall be declared by the Government a "reserved area" under the mining laws of the country or treated in a comparable manner to ensure the rights of the Fund. Such a declaration shall not affect existing mining rights. The Agreement shall specify procedures whereby portions of the reserved area shall be relinquished by the Fund, except for smaller "target areas" of limited size designated as a result of the work of the project. The obligation of the Government to make Replenishment Contributions to the Fund shall relate to all production of natural resources which have been identified in the target areas and reported upon by the Fund project. The basis for Government liability for Replenishment Contributions shall thus be the location of the deposit in a target area defined by the project and reference in a project report to the existence of a resource or resources.

20. Replenishment Contributions from user countries shall be levied at

the uniform rate of 2 percent of the annual value of produced commodities and shall be payable at this rate for a period of 15 years from the start of commercial production or until a ceiling as may be determined in accordance with paragraph 22 below is reached, whichever is earlier. (3)

21. The 15-year period over which the 2 percent shall be due shall be counted from the time the mine commences commercial production and shall be automatically suspended for periods of non-production and extended accordingly. Where economically marginal projects may be prevented from coming into production because of the 2 percent Replenishment Contribution, the Government and the Fund may agree to a lower percentage over a period longer than 15 years or to the payment of a percentage higher than 2 percent for the balance of the period.

22. In principle, a ceiling shall be imposed on repayment contributions to the Revolving Fund. Insufficient experience is available to establish the level at which such a ceiling should be imposed. Such data as are available, however, suggest that the ceiling should probably be in the region of a multiple of 15 times the original investment by the Fund, in constant prices. Accordingly, should a user country in consultation with the Director of the Fund consider that a level of repayments approaching this ceiling is in sight of being reached, the situation shall be reviewed by the Governing Council.

23. All Replenishment Contributions shall be made in convertible currencies.

Funding Policy

24. Funding policies of the Fund should reflect the nature of the Fund's phased operations. Full funding of projects would be impractical and would tie up too large a proportion of the over-all financial resources available, because only a limited number of projects undertaken will carry through to subsequent phases and the costs of succeeding phases generally can be only roughly estimated since their nature and extent are dependent upon the results of preceding phases. The decision not to apply full funding, however, can only be made with the understanding that the level of contributions in future years will not decrease and that further activities on existing projects will have to receive priority over new projects. Thus, in general, new projects would have to be undertaken on the basis of additional contributions.

II. ORGANIZATION AND MANAGEMENT

25. The organization and management of the Fund are designed to ensure maximum efficiency. The Administrator will establish clear

lines of responsibility for the direction of the Fund, while maintaining close relationships between it, UNDP, the United Nations and other concerned international agencies. The Fund will build up its own technical and managerial staff as expeditiously as possible and commensurately with its resources and volume of operations. In the initial stages the Administrator will make all possible use of existing technical staff both in UNDP and in the United Nations, as well as of available administrative machinery and contracting and other facilities, with appropriate arrangements for reimbursement of their costs from the Fund's resources.

Organization

26. In conformity with Economic and Social Council resolution 1762 (LIV), paragraph 1(n), the Administrator, in full consultation with the Secretary-General, appointed a Director of the Fund, and the appointment has been confirmed by the Governing Council.

27. A Joint Operations Group has been established to advise on all major decisions regarding Fund projects. It consists of officers of UNDP, World Bank and the substantive office of the competent agency, which for mineral resources will generally be the United Nations. (4) The Joint Operations Group is chaired by the Director and meets in the composition required for the subject under discussion, including a representative of the UNDP Regional Bureau concerned. It can associate outside experts selected for their professional competence in the technical field covered by the project.

28. The Administrator, in co-operation with the Secretary-General, may from time to time seek on major issues the advice of a panel of experts of international reputation in the field of resources exploration.

29. Project decisions shall be the responsibility of the Administrator, delegated as appropriate to the Director. Major project decisions, including those concerning such matters as the manner of execution of projects, the appointment of project managers, the award of major contracts and the modification, re-orientation or discontinuation of projects, shall be the subject of prior examination by the Joint Operations Group. The Administrator shall take fully into account any recommendations made by the Joint Operations Group on the basis of such examination. Thus the responsibility of the Joint Operations Group for advising on all major project decisions will be integrally related to project management.

30. For each project a project supervisor shall be appointed who will be fully responsible for day-to-day technical management of the project at headquarters. A technical manager shall be designated who will be responsible for the co-ordination and guidance of the project supervisors. Until such time as the Fund has built up its own technical staff, maximum use will be made of the substantive management capability of the United Nations. (5)

31. Similarly, pending the formation of a separate staff, the facilities of UNDP and the United Nations will be used, as appropriate, for the purpose of recruiting project personnel, awarding contracts and purchasing equipment. Procedures will be modelled on those in force for UNDP-financed projects, with such modifications as may be indicated by the operational requirements of the Fund.

Financial Management, Control and Reporting

32. Financial management and control shall be exercised by employing existing staff of UNDP during the initial period. Financial regulations will be worked out for the Fund based upon those in force in UNDP, modified as necessary.

33. Services rendered to the Fund by personnel of UNDP and the co-operating United Nations agencies shall be reimbursed on the basis of additional costs incurred in the preparation and management of projects.

34. In carrying out a project, the Fund shall maintain a separate project account which shall be audited annually by the United Nations Board of Auditors, and shall make available to the user Government, at its request, the auditing report containing information with respect to such account.

Project Requests, Project Agreements and Field Operations

(a) Requests

35. UNDP Resident Representative shall be designated as representatives of the Fund and be the Fund's point of contact in the field. All project requests shall be presented by Governments to the Resident Representatives, who shall forward them to the Fund with their own comments and recommendations.

36. Where review and appraisal of a request or the formulation of a project calls for a field mission, a technical advisor or consultant shall be assigned from existing staff or from outside, and the costs of such a study or project costs if and when a project results from such a mission. The mission shall, in co-operation with the Resident Representative, assist the Government in finalizing its official project request.

(b) Project Agreements

37. Upon authorization by the Administrator, a Project Agreement shall be negotiated with the Government by staff assigned by the Fund under the general supervision of the Resident Representative.

38. Project Agreements shall be standardized as far as possible (see chapter III) but shall in each case specifically describe the goals to be sought and set out the project activities in phases, with the initial phase described in detail.

39. Upon completion of project negotiations with the Government, the project request, accompanied by the recommendation of the Joint Operations Group, shall be forwarded to the Administrator for provisional approval. Projects provisionally approved by the Administrator shall be submitted to the Governing Council for final approval. The proposal of the Administrator shall include the best possible estimate of costs involved, bearing in mind the special characteristics of exploration activities, which may in many cases allow only the initial phase to be costed with any precision.

(c) Field Operations

40. During field operations, the project manager and other project personnel in the field shall keep in close contact with the user Government, the Resident Representative and the Fund.

41. There shall be close user Government/Fund co-operation in the management and operation of individual projects.

42. Locally available professional and other staff, institutional facilities and equipment and supplies shall be utilized by the Fund to the extent conducive to the effective execution of the project. Project personnel shall co-operate with the appropriate professional staff of the Government.

43. In implementing projects financed under the Fund, the Director of the Fund, in co-operation with the user Government, shall give special consideration to recruiting experts, consultants and contractors from developing countries on a preferential basis. In the procurement of equipment and other materials required for Fund projects, the Director shall be invited to purchase suitable and competitive equipment and materials available in developing countries on a preferential basis.

44. The user Government shall participate in the selection of contractors and the recruitment of experts for Fund-financed projects.

45. The Fund shall keep the Government informed periodically of the progress of the project and shall advise the Government immediately of any major developments.

46. Secrecy of information and data produced by and for a project shall be maintained except as the Fund and the user Government may otherwise agree.

47. If upon completion of a project the user Government and the Fund agree that it shall be more economical to sell in the Government's territory rather than to export therefrom any equipment or material imported into such territory by or on behalf of the Fund for the purpose of carrying out a project, the user Government shall use its best efforts to facilitate such sale and shall permit the free transfer of the proceeds thereof outside the user Government's territory.

III. PROJECT AGREEMENTS

48. As the Fund is not covered by the UNDP Basic Agreements with Governments, a separate legal instrument (herein called a Project Agreement) shall have to be agreed upon between the Fund and any Government for which the Fund shall carry out a project of exploration of natural resources under its national jurisdiction. A standard form of such a Project Agreement has been drafted (document No. UNDP/PROG/44 UNDP/PROG/FIELD/60 UNDP/PROG/HQTRS/78), to be adapted in order to meet specific requirements of individual projects. Each user Government shall negotiate with the Fund its own Agreement consistent with the operational procedures and administrative arrangements as set forth in the present document. The "Model" Agreement shall serve only as a model and shall form the basis of negotiations.

49. The Project Agreement is designed to regulate the mutual rights and obligations of the Fund and of a user Government and should be fully consistent with the principle of permanent sovereignty of States over their natural resources as recognized by the relevant resolutions of the United Nations General Assembly. Under the Project Agreement, the Fund shall undertake to carry out a project of exploration of natural resources under the national jurisdiction of the Government while the latter shall undertake to make Replenishment Contributions to the Fund under the circumstances and in the manner described in chapter I.

50. Unlike UNDP, the Fund will not enter into both Basic Agreements and specific Project Documents, but only into a single Project Agreement for each exploration project. (6)

This approach is necessary for the following reasons:

(a) Until more experience is gained, it is desirable to proceed on a flexible case-by-case basis rather than to start by entering into a large number of Basic Agreements with Governments, many of which may not wish or be able to receive Fund assistance in the immediate future;

(b) Some of the elements to be included in the Project Agreement may not be covered by the Government's mining laws or other statutes applicable to the Fund's activities so that the Government will need to obtain specific legislative authorization or ratification of its agreement with the Fund. This kind of authorization or ratification is not likely to be granted on an open-ended basis but will usually be restricted to specific areas and/or to specific natural resources. Thus, a separate authorization or ratification may well be required for each specific project in a particular country so that the purpose of simplification sought by a UNDP type Basic Agreement could not be achieved.

51. As a consequence, the Project Agreement covers the essential elements of both the UNDP Basic Agreement and the UNDP Project Document, but, more importantly, it includes a substantial number of detailed provisions on the Replenishment Contribution obligations as well as various other features not normally found in development

assistance projects concluded within the United Nations system. The Project Agreement includes provisions on matters contained in the following paragraphs 52 to 57.

52. Privileges and immunities: The Fund shall be accorded the same privileges and immunities as are or shall be accorded to UNDP.

53. Settlement of disputes: The Project Agreement shall provide for the settlement by arbitration of any controversy that might arise between the Fund and the Government under the Project Agreement which is not settled by negotiation or other agreed mode of settlement.

54. Work plan: Attached to the Project Agreement is a work plan describing the work to be performed by the Fund in carrying out the project. The work plan and the respective provisions in the Project Agreement shall be different depending on whether the Fund is to carry out an exploration project or a feasibility investigation.

(a) In case of an exploration project, it is difficult to define in advance the kind and the amount of work to be carried out by the Fund as well as the Fund's respective expenditure for the whole project period. The work plan shall therefore include the minimum work which the Fund shall be obligated to carry out as the first phase of the project, expressed in inputs in monetary terms, and in objectives to be agreed between the Fund and the Government in each specific case. The description of minimum work to be carried out should be followed by a broad outline of additional work that based on its findings, the Fund may subsequently undertake.

The Project Agreement shall entitle the Fund, based on an objective technical evaluation of the results obtained in carrying out the exploration project, to decide at any time whether or not further work is justified, and, if so, its nature and extent. Amendments to the work plan of a project to be carried out by the Fund shall be made in consultation with the user Government. Whenever, in the course of the Fund's activities on a project, qualified public or private interests are willing to undertake the further exploration activities leading to exploitation in the Fund's project area, the Fund, at the request of the Government, may agree to discontinue its exploration activities, provided that the Fund retains all its rights to Replenishment Contributions with respect to the exploitation in the project area. The Project Agreement for an exploration project shall only cover the Fund's exploration work; if a feasibility investigation is to be carried out (whether in addition to, or without, a previous exploration project of the Fund), a separate Project Agreement shall be concluded between the Fund and the Government;

(b) In case of a feasibility investigation, the work plan shall describe in detail the work to be performed by the Fund, and the Project Agreement shall provide for the Fund's obligation to carry out the work so described.

55. Final report: At the conclusion of the exploration activities carried

out by the Fund, the Fund shall submit a final report to the Government specifying whether or not minerals were identified. If a mineral deposit is identified, it shall be referred to in the final report as a "reported mineral deposit". A "reported mineral deposit" shall be any or all occurrences of reported mineral or minerals having known or potential economic value that have been investigated in a professional manner to the stage where a possible economic potential can be indicated in respect of composition and possible grade and tonnage of valuable elements or compounds in the deposit. The final report shall contain geological, geochemical, geophysical, drilling and other survey data collected during the exploration work. In particular, the results of elemental analysis on all samples shall be recorded in the final report together with illustrations accurately depicting the type, location and extent of the sampling which may entail pitting, trenching, drilling or other means of access to the deposit as determined by the Fund's technical representative on the project. The Project Agreement will provide that the Government shall have an opportunity to express its views to the Fund:

(a) As to whether all obligations under the Agreement have been fulfilled by the Fund;

(b) As to its acceptance of the reported minerals listed in the final report insofar as their occurrence can be deemed to be of potential economic significance. If the Government should not agree that the report meets these requirements, the Government and the Fund shall consult as to the measures required and, failing agreement on such measures, the matter will be submitted to arbitration.

56. Duration of the Project Agreement: The Project Agreement shall be concluded for a period of 30 years. As a consequence, 30 years after the date of the Agreement, the Agreement itself and all the rights and obligations of the parties thereunder shall terminate, irrespective of whether up to that moment the Fund has received Replenishment Contributions on a discovery for 15 years, for a shorter period of time, or not at all. In addition, the Project Agreement shall provide for termination prior to the end of the 30 year period in the following circumstances:

(a) The Government shall have the right to ask for the Project Agreement to be terminated by arbitration if the Fund has failed to perform its obligations and has not improved its performance within a given period of time;

(b) The Fund shall have the right to ask for the Project Agreement to be terminated by arbitration if circumstances (including the failure of the Government to perform its obligations under the Agreement) arise and continue to exist which, in the judgement of the Fund, interfere with the successful completion of its project;

(c) If the Fund is unable to make a discovery of natural resources and decides to discontinue its exploration activities it shall submit a negative final report to the Government stating its findings and its decision. As of the date of this report, the Project Agreement and all the rights and obligations of the parties thereunder (including the Fund's right to receive Replenishment Contributions if discoveries in the project area should subsequently be made by others) shall automatically terminate;

(d) Whenever the entire Replenishment Contribution due the Fund under the Project Agreement shall have been made to the Fund's satisfaction prior to the expiration of the 30 year period, the Project Agreement and all the rights and obligations of the parties thereunder shall terminate upon dispatch by the Fund to the Government of a notification that these payments have been made.

57. Effectiveness of Project Agreement: Both because of the unusual features of the Project Agreement and because of its long duration, the Fund shall use every precaution to ensure that the Project Agreement becomes a valid and binding obligation of the Government. The Project Agreement shall stipulate that it shall not become effective until the submission of a certificate satisfactory to the Fund, of a competent official of the Government, showing that the Project Agreement has been duly authorized or ratified by, and executed and delivered on behalf of, the Government and is legally binding upon the Government in accordance with its terms. Depending on the Government's laws, the authorization or ratification may involve the issuance of a governmental decree or a ratification decision by the Government's legislative authority or other action.

NOTES

(1) See General Assembly resolution 3167 (XXCIII) and Economic and Social Council resolution 1762 (LIV).

(2) Countries where insufficient geological and related data are available for the purpose of making a request will normally first be investigated through a UNDP project financed from the country's IPF. Where a country's IPF cannot accommodate such preliminary reconnaissance, the Fund may in exceptional circumstances undertake a limited survey in such a way as to identify a potential project area.

(3) The relationship between exploration costs for geothermal energy resources and the value of the energy released would be such as to require a formula for the Replenishment Contribution other than the standard formula referred to above. An alternative formula for geothermal energy, and probably also for petroleum and gas, may be developed by the time the Fund is in a position to enter those fields.

(4) The substantive office in the United Nations is the Centre for Natural Resources, Energy and Transport in the Department of Economic and Social Affairs. If the project falls within the specific competence of IAEA or other United Nations agency, the arrangements will be adapted accordingly.

(5) If the project falls within the specific competence of the IAEA or other United Nations agency, responsibility for technical management may be assigned to the respective agency.

(6) UNDP-assisted development projects are covered by two sets of legal instruments: namely, the so-called "Standard UNDP Basic Agreement with Governments" and the so-called "Project Documents". The first, entered into between UNDP and a Government, does not concern any specific development project but is designed to incorporate the basic conditions under which UNDP will assist the Government to carry out such projects. The second, entered into between UNDP, the Government and an Executing Agency charged with the responsibility for carrying out UNDP assistance to a specific project, sets forth in more detail the respective responsibilities of the parties with regard to the project. The advantage of this approach is that certain basic provisions (e.g. regarding privileges and immunities or settlement of disputes) that often require ratification on behalf of the Government, can be agreed upon once and for all with respect to all UNDP-assisted projects carried out while the Basic Agreement is in force.

Notes

Chapter 3

(1) The act, named after its sponsor, Senator McCarren, which aimed at limiting immigration into the United States.

(2) The Engineers Joint Council is, de facto if not de jure, the Board of Directors of the American Engineering Societies.

Chapter 4

(1) In July 1977, following a discussion of a staff study of minerals and energy in the developing countries, the executive board of the World Bank approved a more active role for the World Bank in this sector through increased lending.

Chapter 5

(1) In late 1977, the feasibility study indicated that profits would be on the modest side, about ten percent. Accordingly, Union Carbide, the American concern involved, withdrew. Japan requested that the time for final decision be postponed by a year.

(2) In the summer of 1977, the Japanese withdrew because of shortage of funds and lack of demand for aluminum.

Chapter 6

(1) As of mid-1978 the financial position of the UNDP reflected a very marked recovery.

(2) Exactly what is excessive? The Revolving Fund offers to do work - highly costly work with price tags for individual projects amounting to as much as three million dollars. The repayment rate of two percent is several times less than that which any competent mining company would consider as a bare minimum to meet the risks of geological exploration, mine development and operation, in particular, and political cross-currents, in general. In effect, it is more than evident that governments which object to a two percent levy are either very naive or, more likely, are looking down other roads - roads along which the welfare of the nations will, on more than one occasion, be sacrificed for the gains of special interests.

Chapter 7

(1) Probably in the neighborhood of 100 dollars today (1978).

(2) Currently (1978) there are sound indications that one of copper's major markets - telephone wire - will soon be invaded by glass fibers. It is evident that sources of glass are inexhaustible.

Chapter 9

(1) Unless otherwise specified, all value figures are in Canadian dollars throughout this chapter.

References

Bartley, M.C. Address to Toronto Branch of the Canadian Institute of Mining and Metallurgical Engineers. Northern Miner (Toronto: September 30, 1976): 5.

Brooks, David B. Minerals: An Expanding or a Dwindling Resource? Ottawa: Canada, Department of Energy, Mines and Resources, Mineral Bulletin MR 134, 1973.

Brown, R.D. "A Pause for Reflection." Canadian Institute of Mining and Metallurgy Bulletin, vol. 69, no. 767 (Montreal: March 1976): 165.

_____. "Now It's Ontario's Turn." Canadian Institute of Mining and Metallurgy Bulletin, vol. 67, no. 747 (Montreal: July 1974): 120-22.

Canadian Yearbook. Ottawa: Ministry of Industry, Trade and Commerce, 1975.

Canadian Mines Yearbook. Northern Miner Press. (Toronto, 1966 and 1969/70).

Carman, John S., ed. Survey of World Iron Ore Resources. New York: United Nations, 1970.

Chabot, James R. "Rebuilding British Columbia's Mineral Industry to Meet Future Needs." Address to British Columbia and Yukon Chamber of Mines Sixty-fifth Annual Meeting. Vancouver: January 11, 1977.

Clarke, W.F., and D.R. Grimes. "Planning and Construction of Leaf Rapids - A New Town with a New Concept." Canadian Institute of Mining and Metallurgy Bulletin, vol. 68, no. 756 (Montreal: April 1975): 84-93.

Coal Age. "Colstrip, Montana: The Modern Company Town" vol. 80, no. 5 (New York: May 1975): 54-57.

Council on Environmental Quality. First Annual Report. Washington, D.C., 1970.

Creighton, Donald. John D. Macdonald: The Young Politician. Toronto: Macmillan, 1956.

Emigh, G.D. "World Phosphate Reserves - Are They Enough? Engineering and Mining Journal, vol. 173, no. 4 (New York: April 1972): 90.

Engineering and Mining Journal. "Must Mining Exploration Die in Canada?" vol. 174, no. 5 (New York, May 1973): 76-79.

Engineering and Mining Journal. "Exploration Dollars Cut Back in British Columbia for Third Consecutive Year," vol. 176, no. 12 (New York: July 1975): 17.

Engineering and Mining Journal. "Saskatchewan Announces Plan to Expropriate All or Part of Potash Industry," vol. 176, no. 12 (New York, December 1975): 23

Engineering and Mining Journal. "Saskatchewan," vol. 179, no. 1 (New York: January 1978): 137-38.

Engineering and Mining Journal. "Market" vol. 176, no. 2 (New York, February 1975): 44.

Howard-Goldsmith, R.C. "The Mineral Picture of Africa." Paper given at the United Nations Seminar on Mineral Economics. Ankara, 1971.

Lowell, J. David. "Copper Resources in 1970." Mining Engineering, vol. 22, no. 4 (New York: April 1970): 68.

London Times. Higher Education Supplement. March 31, 1972.

Meadows, Donella, Dennis L. Meadows, Jorgen Randers, and William B. Behrens, III. The Limits to Growth. New York: Universe Books, 1972.

Mining Journal. "Potash: Final Test for Pro-Rationing," vol. 289, no. 7428 (London: December 30, 1977): 530.

Mining Magazine, vol. 135, no. 3 (London: September 1976): 232-245.

O'Neil, Thomas J. "The Mineral Depletion Allowance. Its Importance to Non-Ferrous Metal Mining." Part I. Mining Engineering, vol. 26, no. 10 (Salt Lake City, Utah: October 1974): 61-64.

Powis, Alfred. Address to Shareholders of Noranda Mines, Limited. Engineering and Mining Journal, vol. 175, no. 6 (New York: June 1974): 32.

"Savage tax policies turn profit into loss at Central Canada Potash," Panorama, vol. 20, no. 9 (Toronto: October 1978): 9.

Skelding, Frank. Small-Scale Mining in the Developing Countries (ST/ECA/155). New York: United Nations, 1972.

Tech Corporation Limited, Report for the Six Months Ended March 31, 1975 (Vancouver, 1975): 3.

Thant, U. "The Decade of Development." Address to the Denmark Students Association. Copenhagen: May 8, 1962 (Press Release SG/1194).

United Nations. Proceedings of the United Nations Conference on the Conservation and Utilization of Resources. Mineral Resources, vol. II. New York, 1951.

U.S. Department of the Interior, Bureau of Mines. Mineral Facts and Problems. Bulletin 650. Washington, D.C. 1970.

Waddingham, D.G. "Royal Bank Economist Sees Need to Rationalize Natural Resource Taxation." Canadian Institute of Mining and Metallurgy Bulletin, vol. 69, no. 765 (Montreal: January 1976): 93.

Bibliography

BIBLIOGRAPHY

Adams, Walter, ed. The Brain Drain. New York: Macmillan, and London: Macmillan, 1968.

Association of Geoscientists for International Development. New Directions in Mineral Development Policies. Report of the Bagauda International Workshop, Bagauda, Nigeria, September 1975.

Bailley, P.A. "The Problems of Converting Resources to Reserves." Mining Engineering (January 1976): 27-37.

Banks, Ferdinand E. The Economics of Natural Resources. New York: Plenum Press, 1976.

Barnett, H.J., and M. Chandler. Scarcity and Growth: The Economics of Natural Resources Availability. Baltimore: Johns Hopkins University Press, 1963.

Berger, A.R., ed. Geoscientists and the Third World: A Collective Critique of Existing Aid Programmes. Paper 74-57, 1975. Ottawa: Geological Survey of Canada.

Bosson, Rex and Bension Varon. The Mining Industry and the Developing Countries. A World Bank Research Publication. Oxford: Oxford University Press, 1977.

Brooks, David B., ed. Resource Economics: Selected Works of Orris C. Herfindahl. Washington, D.C.: Resources for the Future, 1974.

_____, and P.W. Andrews. "Mineral Resources, Economic Growth and World Population," Science, vol. 185, no. 4145, (July 5, 1974).

Canada, Department of Energy, Mines and Resources. Quantitative Impact of Minerals on Canadian Economic Development: A Partial Analysis. Ottawa: Mineral Bulletin MR 173, 1975.

Cole, H.S.D. et al., eds. Models of Doom: A Critique of the Limits to Growth. New York: Universe Books, 1973.

Cranstone, D.A., and H.S. Martin. "Are Ore Discovery Costs Increasing?" Canadian Mining Journal, April 1973.

Drolet, Jean-Paul. Mining Legislation and the Role of Responsible Authorities. Ottawa: Canada, Department of Energy, Mines and Resources, 1975.

English, A.L. "The Changing Nature of Mine Finance in Canada." In Proceedings of the Council of Economics, American Institute of Mining, Metallurgical and Petroleum Engineers (AIME), 1973 pp. 7-14.

Evans, J.B. "How Bill 31 will Affect an Actual Copper Mine," Northern Miner (June 6, 1974).

Fishlow, A., C.F. Diaz-Alejandro, R.R. Fagen, and R.D. Hansen. Rich and Poor Nations in the World Economy. Council on Foreign Relations. New York: McGraw-Hill, 1978.

Girvan, Norman. Corporate Imperialism: Conflict and Expropriation - Transnational Corporations and Economic Nationalism in the Third World. New York: M.E. Sharpe, 1976.

Glasby, G.P., ed. Marine Manganese Deposits. Amsterdam/New York: Elsevier Scientific Publishing Co., 1977.

Golway, M.A. Japanese Involvement in British Columbia Copper. Ottawa: Canada, Department of Energy, Mines and Resources, Mineral Bulletin MR 145, 1975.

Govett, G.J.S., and M.H. Govett, eds. World Mineral Supplies: Assessment and Perspective. Developments in Economic Geology 3. Amsterdam: Elsevier Scientific Publishing Co., 1976.

Guccione, Eugene. "Mine Financing: From Bad to Worse." Mimeographed. Paper presented at the Rocky Mountain Energy Minerals Conference, sponsored by the U.S. Bureau of Land Management, Billings, Montana, October 15-16, 1975.

Hale, W.E., G.J.S. Govett, and M.H. Govett. "Post-Graduate Training and Research in Applied Science in Developing Countries: The Case of Mineral "Exploration" Natural Resources Forum, vol. 1, no. 3 (April 1977): 227-38.

Harkin, Daniel. "Systematic Mineral Exploration." Natural Resources Forum, vol. 1, no. 1 (October 1976): 29-37.

Herfindahl, Orris C. "The Long Run Costs of Minerals." Three Studies in Mineral Economics. Washington, D.C.: Resources for the Future, 1961.

Hoskins, J.R. and W.R. Green. Mineral Industry Costs. Spokane, Washington: Northwest Mining Association, 1977.

McDivitt, James F., and Gerald Manners. Minerals and Men: An Exploration of the World of Minerals and Metals, revised and enlarged. Baltimore: Johns Hopkins University Press, 1974.

McKern, R.B. Multinational Enterprises and Natural Resources. Sydney: McGraw-Hill, 1976.

Mikdashi, Z. The International Politics of Natural Resources. Ithaca, New York: Cornell University Press, 1976.

Mikesell, Raymond. "Rate of Exploitation of Exhaustible Resources: The Case of an Export Economy." Natural Resources Forum, vol. 1, no. 1, (October 1976): 39-46.

_____, and E.H. Conant. Know-How Transfer Through Employee Training: Cases of Natural Resources Industries. New York: Fund for Multinational Management Education, 1977.

_____, ed. Foreign Investment in the Petroleum and Mineral Industries. Baltimore: Johns Hopkins University Press for Resources for the Future, 1971.

_____. Foreign Investment in Copper Mining: Case Studies of Mines in Peru and Papua New Guinea. Baltimore: Johns Hopkins University Press for Resources for the Future, 1975.

Moran, Theodore H. Multinational Corporations and the Politics of Dependence: Copper in Chile. Princeton, New Jersey: Princeton University Press, 1974.

Morton, K. and Peter Tulloch. Trade and Developing Countries. New York: John Wiley and Sons, 1977.

National Commission on Supplies and Shortages. Government and the Nation's Resources. Washington, D.C., December 1976.

Sauvant, Karl and Hajo Hasenpflug. The New International Economic Order, Confrontation or Cooperation between North and South? Boulder, Colorado: Westview Press, 1977.

Shachter, O. Sharing the World Resources. New York: Columbia University Press, 1977.

Seidman, Anne, ed. Natural Resources and National Welfare. New York: Praeger, 1975.

Smith, David, and L. Wells. Negotiating Third World Mineral Agreements. Cambridge, Mass.: Ballinger, 1975.

Sutulov, Alexander. Minerals in World Affairs. Salt Lake City: University of Utah Printing Services, 1972.

Tilton, John E. "The Continuing Debate over the Exhaustion of Non-Fuel Mineral Resources." Natural Resources Forum, vol. 1, no. 3 (January 1977): 167-173.

_____. The Future of Non-Fuel Minerals. Washington, D.C.: Brookings Institution, 1977.

Tucker, R.W. The Inequality of Nations. New York: Basic Books, 1977.

United Nations. Economic Implications of Sea-Bed Mineral Development in the International Area (A/CONF.62/25). New York, 1974.

_____. Recent Advances in Mining and Processing of Low-Grade and Submarginal Mineral Deposits. New York: Pergamon Press, 1976.

_____. Economics of Mineral Engineering. Proceedings of a seminar organized in Ankara, Turkey, 1976. London: Mining Journal Books, 1976.

_____. Negotiation and Drafting of Mining Development Agreements. Proceedings of a workshop organized in Buenos Aires, Argentina, 1973. London: Mining Journal Books, 1976.

_____. Exploration for Natural Resources in Developing Countries: Financial Requirements and Substitutional Mechanisms. Background Paper for the Intergovernmental Group of Experts on Mineral and Energy Exploration (ESA/NRET/AC.12/1) March 7, 1978.

_____. Minerals: Salient Issues. Report of the Secretary-General (E/C.7/68) March 29, 1977.

_____. The Impact of Multinational Corporations on Development and on International Relations (55/rev.1.ST/ESA/6). New York, 1974.

_____. The Impact on Multinational Corporations on Development and on International Relations. Technical Papers: Taxation. (ST/ESA/11). New York, 1974.

_____. Interregional Seminar on the Application of Advanced Mining Technology. Proceedings Ottawa, Canada, May 21, June 3, 1973 (DP/UN/INT-72-064). New York, 1975.

_____. Seminar on Mining Legislation and Administration, Manila, 1969, Proceedings (E/CN 11/I and NR/L 90). New York, November 17, 1969.

_____. Problems of Raw Materials and Development (Center for Economic and Social Information). Declaration and program of action adopted by the Sixth Special Session (ES1.E21) May 12, 1974.

Vogeley, W.A., ed. Economics of the Mineral Industries. New York: American Institute of Mining, Metallurgical and Petroleum Engineers (AIME), 1976.

Waelde, Thomas. "Lifting the Veil from Transnational Contracts A Review of Recent Literature." Natural Resources Forum, vol. 1, no. 3 (April 1977): 277-84.

Walthier, Thomas N. "Problems of Foreign Investment in Natural Resources." Paper presented at the International Minerals Acquisition and Operations Institute, Denver, Colorado, October 1974.

Weintraub, A., E. Schwartz, and P.R. Aronson, Eds. The Economic Growth Controversy. White Plains, N.Y.: International Arts and Sciences Press, 1973.

Zorn, Stephen. Renegotiating Mining Agreements: The Case of Bougainville Copper. United Nations Workshop on Negotiations with Foreign Investors, New Delhi, 1975.

___. "New Developments in Third World Mining Agreements." Natural Resources Forum, vol. 1, no. 3 (April 1977): 239-50.

Zwartendyk, Jan. "Problems of Interpretation of Data on Mineral Resources, Production and Consumption." Natural Resources Forum, vol. 1, no. 1 (October 1976): 7-15.

Index

About the Author and the Editor

THE AUTHOR

JOHN CARMAN is a mining engineer with 30 years experience covering private industry and government in Canada and with the United Nations. He received his B.Sc. from Michigan Technological University and his Ph.D. from Columbia University. He worked in various capacities in Canada, his native country, as a miner, shift boss, mine superintendent, manager of an exploration company seeking uranium, and vice-president of a uranium-producing enterprise. In addition, he was retained by the Royal Commission on Canada's Economic Prospects to estimate the base metal reserves of that country and commissioned by the Canadian Goverbnment under the Colombo Plan to work out the development plan for the Kiriburu iron ore deposit in India. During the fifties, he also did consulting work in the United States, Canada, Haiti, Ireland, Brazil, and Bolivia, the latter for the United Nations.

In 1958, he joined that organization and, over the next two decades, he participated actively in all facets of United Nations activities in mineral exploration and advised governments frequently on mineral policy. He undertook missions to about 40 countries during this period, varying in duration from a few days to as much as two years.

Before leaving the United Nations in 1976, he served as Technical Manager of the Revolving Fund for Natural Resources Exploration. A recipient of various honors during his scholastic work, Carman was awarded the Orden al Merito Bernardo O'Higgins (Chile) in 1961 and the Orden de Mayo al Merito en grado de Comendador (Argentina) in 1969 but, because of United Nations regulations, he was unable to accept. He also received the Doctor of Science degree, honoris causa, from the University of New Brunswick in 1972 and the Board of Control Silver Medal of Michigan Technological University in 1973.

John Carman has lectured extensively in universities and professional meetings in both North America and abroad. He is the author of a dozen publications and the editor of the mammoth Survey of World Iron Ore Resources published by the United Nations in 1970.

THE EDITOR

Bension Varon is currently a senior economist at the World Bank where he has served in various capacities since 1965. During 1975-77, while on leave from the World Bank, he was Assistant Director in charge of policies and projections at the United Nations Center for Natural Resources, Energy and Transport. Mr. Varon has written extensively on natural resources and trade and is a former editor of the Natural Resources Forum, a quarterly publication of the United Nations. He is a citizen of Turkey and holds a Ph.D. in Business and Applied Economics from the University of Pennsylvania.

Pergamon Policy Studies

No. 1 Laszlo—*The Objectives of the New International Economic Order*
No. 2 Link/Feld—*The New Nationalism*
No. 3 Ways—*The Future of Business*
No. 4 Davis—*Managing and Organizing Multinational Corporations*
No. 5 Volgyes—*The Peasantry of Eastern Europe, Volume One*
No. 6 Volgyes—*The Peasantry of Eastern Europe, Volume Two*
No. 7 Hahn/Pfaltzgraff—*The Atlantic Community in Crisis*
No. 8 Renninger—*Multinational Cooperation for Development in West Africa*
No. 9 Stepanek—*Bangledesh—Equitable Growth?*
No. 10 Foreign Affairs—*America and the World 1978*
No. 11 Goodman/Love—*Management of Development Projects*
No. 12 Weinstein—*Bureacratic Opposition*
No. 13 De Volpi—*Proliferation, Plutonium, and Policy*
No. 14 Francisco/Laird/Laird—*The Political Economy of Collectivized Agriculture*
No. 15 Godet—*The Crisis in Forecasting and the Emergence of the "Prospective" Approach*
No. 16 Golany—*Arid Zone Settlement Planning*
No. 17 Perry/Kraemer—*Technological Innovation in American Local Governments*
No. 18 Carman—*Obstacles to Mineral Development*
No. 19 Demir—*Arab Development Funds in the Middle East*
No. 20 Kahan/Ruble—*Industrial Labor in the U.S.S.R.*
No. 21 Meagher—*An International Redistribution of Wealth and Power*
No. 22 Thomas/Wionczek—*Integration of Science and Technology With Development*
No. 23 Mushkin/Dunlop—*Health: What Is It Worth?*
No. 24 Abouchar—*Economic Evaluation of Soviet Socialism*
No. 25 Amos—*Arab-Israeli Military/Political Relations*
No. 26 Geismar/Geismar—*Families in an Urban Mold*
No. 27 Leitenberg/Sheffer—*Great Power Intervention in the Middle East*
No. 28 O'Brien/Marcus—*Crime and Justice in America*
No. 29 Gartner—*Consumer Education in the Human Services*
No. 30 Diwan/Livingston—*Alternative Development Strategies and Appropriate Technology*
No. 31 Freedman—*World Politics and the Arab-Israeli* Conflict
No. 32 Williams/Deese—*Nuclear Nonproliferatrion*
No. 33 Close—*Europe Without Defense?*
No. 34 Brown—*Disaster Preparedness*
No. 35 Grieves—*Transnationalism in Politics and Business*

No. 36 Franko/Seiber—*Developing Country Debt*
No. 37 Dismukes—*Soviet Naval Diplomacy*
No. 38 Morgan—*The Role of U.S. Universities in Science and Technology for Development*
No. 39 Chou/Harmon—*Critical Food Issues of the Eighties*
No. 40 Hall—*Ethnic Autonomy—Comparative Dynamics*
No. 41 Savitch—*Urban Policy and the Exterior City*
No. 42 Morris—*Measuring the Condition of the World's Poor*
No. 43 Katsenelinboigen—*Soviet Economic Thought and Political Power in the USSR*
No. 44 McCagg/Silver—*Soviet Asian Ethnic Frontiers*
No. 45 Carter/Hill—*The Criminal's Image of the City*
No. 46 Fallenbuchl/McMillan—*Partners in East-West Economic Relations*
No. 47 Liebling—*U.S. Corporate Profitability*
No. 48 Volgyes/Lonsdale—*Process of Rural Transformation*
No. 49 Ra'anan—*Ethnic Resurgence in Modern Democratic States*